Shmuel Friedland, Mohsen Aliabadi
Analysis and Probability on Graphs

Also of Interest

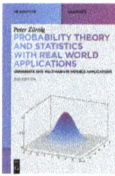

Probability Theory and Statistics with Real World Applications
Univariate and Multivariate Models Applications
2nd Edition
Peter Zörnig, 2024
ISBN 978-3-11-133220-8, e-ISBN (PDF) 978-3-11-133227-7,
e-ISBN (EPUB) 978-3-11-133232-1

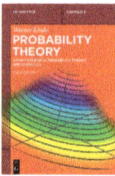

Probability Theory
A First Course in Probability Theory and Statistics
2nd Edition
Werner Linde, 2024
ISBN 978-3-11-132484-5, e-ISBN (PDF) 978-3-11-132506-4,
e-ISBN (EPUB) 978-3-11-132517-0

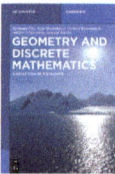

Geometry and Discrete Mathematics
A Selection of Highlights
2nd Edition
Annika Schürenberg, Gerhard Rosenberger, Anja Moldenhauer,
Benjamin Fine, Leonard Wienke, 2022
ISBN 978-3-11-074077-6, e-ISBN (PDF) 978-3-11-074078-3,
e-ISBN (EPUB) 978-3-11-074093-6

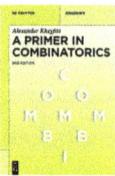

A Primer in Combinatorics
2nd Edition
Alexander Kheyfits, 2021
ISBN 978-3-11-075117-8, e-ISBN (PDF) 978-3-11-075118-5,
e-ISBN (EPUB) 978-3-11-075124-6

Shmuel Friedland and Mohsen Aliabadi

Analysis and Probability on Graphs

—

DE GRUYTER

Mathematics Subject Classification 2020
Primary: 60J05, 60J20, 05C05; Secondary: 15A18

Authors

Prof. Shmuel Friedland
University of Illinois at Chicago
Department of Mathematics
851 South Morgan Street
Chicago
Illinois, 60607
U.S.A.
friedlan@uic.edu

Prof. Mohsen Aliabadi
University of California, San Diego
Department of Mathematics
9500 Gilman Drive
La Jolla
California, 92093
U.S.A.
maliabadisr@ucsd.edu

ISBN 978-3-11-133692-3
e-ISBN (PDF) 978-3-11-133738-8
e-ISBN (EPUB) 978-3-11-133809-5

Library of Congress Control Number: 2024947902

Bibliographic information published by the Deutsche Nationalbibliothek
The Deutsche Nationalbibliothek lists this publication in the Deutsche Nationalbibliografie;
detailed bibliographic data are available on the Internet at http://dnb.dnb.de.

© 2025 Walter de Gruyter GmbH, Berlin/Boston
Cover image: berya113 / iStock / Getty Images Plus
Typesetting: VTeX UAB, Lithuania

www.degruyter.com
Questions about General Product Safety Regulation:
productsafety@degruyterbrill.com

Preface

Analysis and Probability on Graphs, abbreviated as APG, forms a foundational framework across various advanced mathematical topics. It serves as a critical tool in disciplines such as computer science, physics, engineering, bioinformatics, economics, and social sciences, particularly in analyzing the dynamics of large-scale events on graphs. Many of these dynamic phenomena can be modeled through appropriate probabilistic processes applied to large graphs. One real-world application of combining graph theory with analysis and probability is the *social network scenario*: Consider a social network where users are connected based on their interactions or shared interests. The analysis of these connections through graph theory, combined with probability, can produce personalized recommendations. From graph theory perspective, represent users as nodes in a graph and connect users with edges if they have interacted or shared common interests. From probability perspective, assign probabilities to edges based on the strength or frequency of interactions between connected users. Analyze the graph to identify communities or groups of users who exhibit similar behavior or preferences. When a user expresses interest in a particular item (such as a movie, book, or product), the system may predict the likelihood of interest from their connected peers. Calculate the probability that the user will like the item based on the preferences of their connected neighbors in the graph. Recommend items with the highest probability of being liked by the user, considering the preferences of their network connections. Adjust recommendations dynamically as the user's interactions and the graph evolve.

Our goals. In our discussions, we prioritize understanding the motivation of each statement over getting immersed in details. Our approach is to aim for rigor where possible, but in other cases, we focus on conveying the concept of each statement. The main goal is to explore and explain various topics in this direction and provide a background for advanced undergraduate and graduate students in mathematics, physics, and computer science. The book is designed to be accessible to individuals with a basic understanding of graph theory and probability theory, making it suitable for upper-level undergraduates and those involved in independent study, using it as a reference, or exploring research with a basic background in probability and graph theory. The book may also be used as a textbook. Covering all these topics in a one-semester graduate course could be challenging. Nevertheless, since many sections of the book are independent, instructors have the flexibility to select and incorporate relevant sections based on their specific needs.

Organization. From historical perspective, this textbook originates from years of teaching courses in probability theory, graph theory, combinatorics, and matrix theory by the first author at the University of Illinois, Chicago. The book has seven chapters. The first four chapters may be used as a textbook for advanced undergraduate courses related to probability theory, random walks, and probability on graphs in mathematics and mathematical computer science departments. The materials covered in these chapters may be used to investigate real-world problems. The last three chapters

https://doi.org/10.1515/9783111337388-201

are intended for graduate students as a reference book or for individual study or research.

We provide a brief overview of its contents. Chapter 1 focuses on fundamental concepts in probability theory, covering aspects such as probability space, random variables, expected values, and moments. Theorems related to assessing the probability of a random variable, including Markov's inequality and Chebyshev's inequality, are presented. Certain important random variables such as Bernoulli, binomial, and Poisson random variables and their properties are discussed. Chapter 2 is devoted to undirected and directed graphs. We first review the basic notions of undirected graphs. We explore walks, trails, and paths in undirected graphs, as well as examine these concepts in trees and bipartite graphs, along with associated results. Subsequently, we classify Eulerian circuits. We introduce the basic notions of digraphs (directed graphs) and shift our focus to connectedness in digraphs. Finally, we classify strongly connected digraphs by leveraging tools from number theory, such as results concerning Frobenius numbers. In Chapter 3, we discuss two commonly used random graph models, Gilbert's and Erdös–Rényi's models. We scrutinize the implications of these two models for graph connectivity analysis. We address the k-clique property of graphs, a tool for identifying the "dense" subgraphs within a given graph. We also study isolated vertices and connectivity, investigating threshold functions through the Poisson distribution. In Chapter 4, we introduce adjacency matrices for both undirected and directed graphs, facilitating the examination of various graph properties such as vertex degrees, walk enumeration, identification of odd cycles, and connectivity analysis. We also cover block matrices in the context of strong connectivity in directed graphs, alongside discussions on reduced graphs and reducible matrices. In Chapter 5, we explore Markov chains: what they are, how to simulate them, and how they behave on graphs. We analyze stationary distributions and their properties, along with spectral properties of matrices related to Markov chains. We cover reversible Markov chains and then move on to a short discussion of the Perron–Frobenius theorem. We conclude the chapter with a discussion of the mean first passage time, mean recurrence time, and the Kemeny constant of a Markov chain. Chapter 6 links symbolic dynamics with graph theory by interpreting sequences as graph walks, exemplified by a Fibonacci sequence. It covers hard-core configurations, their recurrence relations, and the **MCMC** algorithm. The Shannon capacity for a digraph G with cycles is introduced, related to the spectral radius of its adjacency matrix. The chapter also defines entropy $H(\mu)$ for a distribution μ, explores the pressure function and its properties, and a pressure gradient simulation.

In Chapter 7, on a subshift of finite type, we introduce a pseudo-metric defined by a nonnegative matrix that satisfies the cycle condition. We determine the exact value of the Hausdorff dimension for a G^∞, induced by a digraph G, with respect to this specific pseudo-metric. We also evaluate the Hausdorff dimension of the limit set of a finitely generated free group of isometries acting on a locally finite tree.

We include a brief overview of topics related to basic set theory, mathematical reasoning, basic abstract algebra, mathematical analysis and topology, and introductory linear algebra in our appendix.

Regarding the solved and unsolved problems, we can classify them into two categories:

- Solved problems are labeled as *Worked-out Problem a.b-c*, where a denotes the chapter number, b represents the section number, and c corresponds to the problem number.
- Unsolved problems are labeled as *Problem a.b-c*, where a denotes the chapter number, b represents the section number, and c corresponds to the problem number.

Features. We highlight the following unique features of the book:

- *Motivational Discussions:* The book features discussions elucidating the motivations behind fundamental concepts.
- *Problems:* All chapters include many examples, solved and unsolved problems, which help students apply what they have learned practically.
- *Comprehensive Coverage:* The textbook goes beyond conventional topics, exploring areas rarely covered in probability on graph-related literature. Specifically, we have compiled subjects recognized by the community but never previously assembled in a single book.

Feedback. We hope that any lingering errors and typographical oversights are not excessively held against us, and we sincerely hope they do not present a significant inconvenience to the reader. It is through the generous assistance of our students and peers that we have minimized these issues to their current extent. We appreciate and encourage corrections, suggestions, and comments, covering areas such as topics, result attributions, updates, typographical errors, etc. Kindly share these with us at

"friedlan@uic.edu" or "maliabadisr@ucsd.edu"

Additionally, we extend our apologies in advance for any omitted references; please notify us of the correct citations.

Fall 2024   Shmuel Friedland and Mohsen Aliabadi

Acknowledgments

We extend our gratitude to those who have contributed to the creation of this book. We appreciate the students of the first author who participated in courses built upon initial iterations of this book, which were originally lecture notes. Their active participation played a key role in detecting and correcting numerous typographical errors.

https://doi.org/10.1515/9783111337388-202

Contents

1 Probability

Historical note. The mathematical theory of probability, as we know it today, has its origins in the study of games of chance. Prominent figures such as Gerolamo Cardano (Figure 1.1), Pierre de Fermat (Figure 1.2), and Blaise Pascal (Figure 1.3) made early contributions in the sixteenth and seventeenth centuries. Christiaan Huygens published a book on the subject in 1657. In the nineteenth century, Pierre Laplace's work laid the foundation for what is now considered the classical definition of probability.

Probability theory primarily focused on discrete events and used combinatorial methods. However, as the field developed, it needed to incorporate continuous variables due to analytical considerations. This evolution led to modern probability theory with Andrey Nikolaevich Kolmogorov, who introduced the concept of a *sample space*, building on ideas from Richard von Mises and incorporating *measure theory*. His axiom

Figure 1.1: Gerolamo Cardano (1501–1576). Photo credit: 17th-century portrait engraving of Cardano. Source: https://en.wikipedia.org/wiki/Gerolamo_Cardano.

Figure 1.2: Pierre de Fermat (1607–1665). Photo credit: Painting by unknown author. Source: https://en. wikipedia.org/wiki/Pierre_de_Fermat.

https://doi.org/10.1515/9783111337388-001

Figure 1.3: Blaise Pascal (1623–1662). Photo credit: Portrait of Pascal in 1691. Source: https://en.wikipedia.org/wiki/Blaise_Pascal.

system for probability theory in 1933 became widely recognized as the basis of modern probability theory. Although this system is dominant, alternative approaches also exist, such as Bruno de Finetti's adoption of finite additivity over countable additivity.

Outline. This chapter unfolds in the following manner: We commence by elucidating the concept of a probability space and investigating its counterparts, including the sample space, event space, and probability function. We explore the properties inherent to probability spaces. Subsequently, we introduce the conditional and independent probabilities. Leveraging the framework of a probability space, we navigate through the intricacies of random variables and their associated concepts, such as the expected value function, moments, variance, and standard deviation, all within the context of a single random variable.

Our discussion extends to encompass multiple random variables, leading us to the notion of covariance and its associated properties. We introduce Bernoulli random variables, using them as a gateway to discuss the inclusion–exclusion principle. Our journey continues with exploring Bernoulli, binomial, and Poisson random variables. We conclude this chapter by discussing the inclusion–exclusion principle and Bonferroni inequalities.

Warm up. Many actions yield outcomes that are largely unpredictable in advance; a simple example is coin tossing under various conditions and repetitions. Probability theory focuses on such actions and their consequences. The mathematical theory starts with the concept of an experiment, which represents an action with uncertain outcomes. This experiment is then redefined as a mathematical construct known as a *probability space*. In broad terms, a probability space associated with a given experiment comprises three essential components.

First, the set consisting of all conceivable outcomes of a specific action. Second, a list of potential events that may occur as a consequence of that action. Third, an assessment of the probabilities associated with these events. For example, if the experiment involves tossing a fair coin twice in a row, then the probability space has the following components. (T representing tail and H representing head. Additionally, an ordered pair (X, Y) signifies that X occurs in the first attempt, and Y occurs in the second attempt.)

- $\{(T, T), (H, H), (T, H), (H, T)\}$, the set of all possible outcomes.
- A list of events includes scenarios such as obtaining heads at least once or having both tosses result in heads.
- Each of $(T, T), (H, H), (T, H)$, and (H, T) is equally likely to be the result of the toss.

1.1 Probability spaces

A *probability space*, often referred to as a probability triple and denoted as $(\Omega, \mathcal{F}, \mathbf{Pr})$, serves as a base for modeling random experiments. As $(\Omega, \mathcal{F}, \mathbf{Pr})$ suggests, a probability space consists of three counterparts: Ω represents the *sample space*, $\mathcal{F}$ stands for the *event space*, and $\mathbf{Pr}$ represents the *probability measure*. In what follows, we define each counterpart. The sample space Ω is a set of all possible outcomes $\omega \in \Omega$ in some random experiment. We assume that Ω is a countable set. It includes the *empty set*, denoted by $\emptyset$, and also Ω. A subset $A \subset \Omega$ is called an *event*. An event space is a specific set of events. More specifically, an event space $\mathcal{F}$ is a set of events satisfying the following conditions:

1. $\mathcal{F} \neq \emptyset$,
2. if $A \in \mathcal{F}$, then $A^c \in \mathcal{F}$,
3. if $A_1, A_2, \ldots, \in \mathcal{F}$, then $\bigcup_{i=1}^{\infty} A_i \in \mathcal{F}$.

If we denote by 2^{Ω} the set of all subsets of Ω, then an event space $\mathcal{F}$ is a subset of 2^{Ω}.

Example 1.1. Let us consider the experiment of tossing a fair coin twice in a row. Then $\Omega = \{(T, T), (H, H), (T, H), (H, T)\}$ is our sample space. We can represent the situation in which both tosses result in the same outcome, either heads or tails, as the subset $A = \{(T, T), (H, H)\}$, which is an event. The corresponding event space is $\mathcal{F} = \{\emptyset, \{(T, T)\}, \{(H, H)\}, \{(T, T), (H, H)\}\}$.

We have defined the sample space Ω and the event space $\mathcal{F}$ for a given experiment, but we have yet to determine the probabilities associated with the experimental outcomes, which is the third counterpart of a probability space. This is accomplished through a mapping known as the *probability measure*, which assigns probabilities to each event in $\mathcal{F}$. A probability measure (or probability distribution) on $(\Omega, \mathcal{F})$ is a function $\mathbf{Pr} : \mathcal{F} \to \mathbb{R}$ satisfying the following conditions:

1. $\mathbf{Pr}(A) \geq 0$ for every $A \in \mathcal{F}$,
2. $\mathbf{Pr}(\Omega) = 1$ (*normalization condition*),
3. $\mathbf{Pr}$ is *countably additive*. That is, $\mathbf{Pr}(\bigcup_{i=1}^{\infty} A_i) = \sum_{i=1}^{\infty} \mathbf{Pr}(A_i)$, provided that $A_1, A_2, \ldots \in \mathcal{F}$ are disjoint.

A probability space is a triple $(\Omega, \mathcal{F}, \mathbf{Pr})$, where Ω is a nonempty set, $\mathcal{F}$ is an event space on Ω, and $\mathbf{Pr}$ is a probability measure on $(\Omega, \mathcal{F})$.

Now for each element $\omega \in \Omega$, we assign a probability value denoted as $p(\omega)$, where $p(\omega) = \mathbf{Pr}(\{\omega\})$. Then $p(\omega)$ is greater than or equal to 0. The normalization condition is that the sum of all the probabilities over the entire sample space, represented as $\sum_{\omega \in \Omega} p(\omega)$, must equal 1. In other words, the total probability mass across the entire sample space is 1. By countable additivity we can express the probability of an event $A \in \mathcal{F}$, denoted as $\mathbf{Pr}(A)$, as the sum of the probabilities for all the sample points within A: $\mathbf{Pr}(A) := \sum_{\omega \in A} p(\omega)$.

Example 1.2. Assume that $\Omega = \{\omega_1, \ldots, \omega_n\}$ is a finite sample space. Let $\mathcal{F} = 2^\Omega$, and define $\mathbf{Pr}(A) = \frac{|A|}{|\Omega|}$ for every $A \in \mathcal{F}$ with $|A| = 0$ if and only if $A = \emptyset$. Such a probability is called *the uniform distribution*.

Let $A, B \in 2^\Omega$ be two events. The *intersection* $A \cap B$ of A and B is the set that consists of all elements that belong to A and B. The *union* $A \cup B$ of A and B is the set that consists of all elements that belong either to A or to B. Then the following relation computing $\mathbf{Pr}(A \cup B)$ is called the *probabilistic principle of inclusion and exclusion*, stated and proved in Proposition 1.25:

$$\mathbf{Pr}(A \cup B) = \mathbf{Pr}(A) + \mathbf{Pr}(B) - \mathbf{Pr}(A \cap B).$$

The *complement* of A in Ω is $A^c := \Omega \backslash A$, which consists of all points in Ω that are not in A. So

$$A \cup A^c = \Omega, \quad A \cap A^c = \emptyset \Rightarrow 1 = \mathbf{Pr}(A) + \mathbf{Pr}(A^c).$$

Given a probability space $(\Omega, \mathcal{F}, \mathbf{Pr})$, some straightforward properties following from the definition are left to the reader for a proof:

- if $A, B \in \mathcal{F}$, then $A \setminus B \in \mathcal{F}$;
- if $A_1, A_2, \ldots \in \mathcal{F}$, then $\bigcap_{i=1}^{\infty} A_i \in \mathcal{F}$;
- if $A, B \in \mathcal{F}$ with $A \subset B$, then $\mathbf{Pr}(A) \leq \mathbf{Pr}(B)$.

We often have partial information about the outcome of an experiment and wish to adjust our beliefs about the outcome based on this information. We are interested in the probability of occurrence of an event A, given that event B has occurred. This updated probability is referred to as *conditional probability*.

For example, let us consider tossing a fair coin three times. First, we want to find the probability of observing exactly one head. Then we can ask what is the probability of observing at least two heads given that we have already observed at least one head. This second situation involves conditional probability. We will start by defining and formulating conditional probability and then revisit the above example.

Let $A, B \subset \Omega$. Then $A|B$ is the *conditional event* that A will occur if B already occurred. This is equivalent to the event $A \cap B|B$. Assume that $\mathbf{Pr}(B) > 0$. Then $\mathbf{Pr}(A|B)$ is the *conditional probability* of the conditional event $A|B$. Thus $\mathbf{Pr}(A|B) = \frac{\mathbf{Pr}(A \cap B)}{\mathbf{Pr}(B)}$.

Notice that:

- the condition $\mathbf{Pr}(B) > 0$ is necessary to ensure that the division is well-defined or, equivalently, that $\mathbf{Pr}(A|B)$ is a meaningful quantity.
- if $\mathbf{Pr}(B) = 0$, then the event B can never occur, making the statement $A|B$ meaningless.
- the definition of conditional probability is often called the *product rule*.

We now revisit our example concerning conditional probability.

Example 1.3. Consider tossing a fair coin three times. We find the probability of observing at least two heads, given that we have already observed at least one head. First, let B be the event of observing at least one head, and let A be the event of observing at least two heads. Clearly, our sample space S has 8 elements. Therefore since $B = S \setminus \{(T, T, T)\}$, we have $\mathbf{Pr}(B) = \frac{7}{8}$. On the other hand, $A = \{(H, H, T), (H, T, H), (T, H, H), (H, H, H)\}$, suggesting $\mathbf{Pr}(A) = \frac{4}{8} = \frac{1}{2}$. Thus $\mathbf{Pr}(A|B) = \frac{\mathbf{Pr}(A \cap B)}{\mathbf{Pr}(B)} = \frac{\mathbf{Pr}(A)}{\mathbf{Pr}(B)} = \frac{4}{7}$.

In certain situations the information from one event may not provide any insights or influence our knowledge of another event. This means that the probability of event A given event B is equal to the probability of event A alone, $\mathbf{Pr}(A|B) = \mathbf{Pr}(A)$. In such cases, we describe events A and B as being *conditionally independent*, that is, events A and B are called independent if $\mathbf{Pr}(A \cap B) = \mathbf{Pr}(A) \cdot \mathbf{Pr}(B)$. It is important to note that this latter definition accommodates the cases where the probability of an event B is equal to zero ($\mathbf{Pr}(B) = 0$), making it a more general formulation. Notice that A and B are called *dependent* if they are not conditionally independent.

Example 1.4. When rolling a die twice, we can calculate the probability of obtaining two fives. Let us define the event A as the occurrence of rolling a five on the first attempt and event B as the occurrence of rolling a five on the second attempt. It is evident that the outcomes of the first and second rolls do not influence each other. Thus the events A and B are independent. On the other hand, the event of getting two fives can also be viewed as the combined event of rolling a five on both the first and second attempts, $A \cap B$. Therefore $\mathbf{Pr}(A \cap B) = \mathbf{Pr}(A) \cdot \mathbf{Pr}(B) = \frac{1}{6} \cdot \frac{1}{6} = \frac{1}{36}$.

1.2 One random variable and related concepts

Random variables. A *random variable*, typically denoted as X, is a function that maps to numerical results representing potential outcomes of a random event or phenomenon. It can be viewed as a variable whose values are contingent on the outcome of an unpredictable event In what follows, we provide a more formal definition of a random variable. Let $(\Omega, \mathcal{F}, \mathbf{Pr})$ be a probability space. Then every function $X : \Omega \to \mathbb{R}$ is called a random variable. This means that for each individual outcome ω within the sample space Ω, the function X assigns a corresponding real number $X(\omega)$. The reason we call X is a random variable is as follows:

Suppose $\Omega = \{H, T\}$ is the sample space representing the outcome of a coin toss. Let us assume that $X(H) = 1$ and $X(T) = -1$. In other words, if "head" appears, then you receive \$1, and if "tail" appears, then you lose \$1. Therefore X qualifies as a random variable, as the outcome of the coin toss remains unknown until the coin is tossed. If the coin is fair, then the probability that X equals 1 ($\mathbf{Pr}(X = 1)$) is 0.5, and the probability that X equals -1 ($\mathbf{Pr}(X = -1)$) is also 0.5.

A *discrete random variable* is one that can take on only countably many values; otherwise, it is called *continuous*. Throughout this textbook, our focus will primarily be on discrete random variables, unless otherwise specified. Additionally, sometimes it would be convenient to consider a more general random variable $X : \Omega \to \Theta$, where Θ is some set, not necessarily the set of real numbers. However, in this textbook, we make the assumption that the codomain of all random variables is the set of real numbers unless explicitly stated otherwise. We will denote the range of X by R_X.

Example 1.5. Let us consider a scenario where a coin is flipped twice, defining the sample space $\Omega = \{(H, H), (H, T), (T, H), (T, T)\}$. We use the random variable X to denote the count of tails that may occur. Each $\omega \in \Omega$ in this context can be associated with a numerical value. For example, $X(\omega_1) = 1$ if $\omega_1 = (H, T)$ and $X(\omega_2) = 2$ if $\omega_2 = (T, T)$. We can conclude that X is a random variable.

Cumulative distribution functions. All the information about the random variable $X : \Omega \to \mathbb{R}$ is stored in the *cumulative distribution function* abbreviated as CDF $F_X : \mathbb{R} \to [0, 1]$:

$$F_X(t) = \mathbf{Pr}(X \le t) = \sum_{X(\omega) \le t} p(\omega),$$

where $\mathbf{Pr}(X \le t)$ is a standard abbreviation for $\mathbf{Pr}(X^{-1}(-\infty, t])$.

CDF has the following properties:
- $F_X(t)$ is a nondecreasing function, i. e., $F_X(t_1) \le F_X(t_2)$ for any $t_1 \le t_2$.
- $\lim_{t \to -\infty} F_X(t) = 0$.
- $\lim_{t \to \infty} F_X(t) = 1$.
- F_X is continuous from the right: $\lim_{t \searrow x} F_X(t) = F_X(x) \, \text{forall} x \in \mathbb{R}$, where $t \searrow x$ means that t approaches x from the *right*, i. e., $t > x$, and t converges to x.

Let us assume that the range of a given random variable X is countable; $R_X = \{x_1, x_2, \dots\}$ with $x_1 < x_2 < \cdots$. As we mentioned previously, we have $\lim_{t \to -\infty} F_X(t) = 0$. Then $F_X(t)$ jumps at each point in the range. In particular, the CDF stays flat between x_i and x_{i+1}, so we can write

$$F_X(t) = F_X(x_i)$$

for $x_i \le t < x_{i+1}$. The CDF jumps at each x_i, and in particular, we can write that for every small enough $e > 0$,

$$F_X(x_i) - F_X(x_i - \epsilon) = \mathbf{Pr}(X^{-1}(x_i)).$$

Therefore, as we listed above, CDF is always a nondecreasing function. Moreover, when t is large enough, $F_X(t)$ approaches 1, which is evident from the fact that the total probability of the sample space, $\mathbf{Pr}(\Omega)$, is equal to 1.

Example 1.6. When tossing a coin twice, let X be the count of observed heads. We would like to determine the CDF of X. Clearly, the range of X is $\{0, 1, 2\}$, $\mathbf{Pr}(X = 0) = \frac{1}{4}$, $\mathbf{Pr}(X = 1) = \frac{1}{2}$, and $\mathbf{Pr}(X = 2) = \frac{1}{4}$. Therefore $F_X(t)$ is given by Figure 1.4:

$$F_X(t) = \begin{cases} 0 & \text{if } t < 0, \\ \frac{1}{4} & \text{if } 0 \le t < 1, \\ \frac{3}{4} & \text{if } 1 \le t < 2, \\ 1 & \text{if } t \ge 2. \end{cases}$$

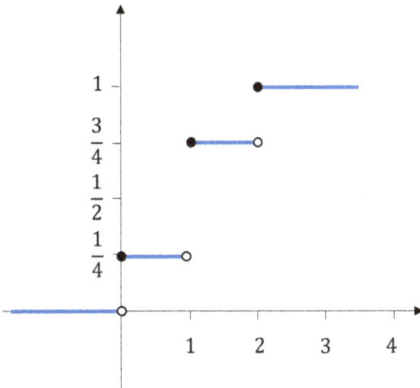

Figure 1.4: Cumulative distribution function.

Remark. Assume that Ω is a *general* sample space, i. e., not necessarily countable. Then a random variable $X : \Omega \to \mathbb{R}$ (satisfying the condition of being a measurable map) is called *countable* if $X(\Omega)$ is a countable set in $\mathbb{R}$. Then the treatment of X is equivalent to the treatment of a random variable on a countable space. This is achieved by identifying all points in Ω whose images under X are identical. Then we obtain the countable space Ω' and $X' : \Omega' \to \mathbb{R}$ such that $F_X = F_{X'}$.

Expected values and Markov's inequality. Given a probability space $(\Omega, \mathcal{F}, \mathbf{Pr})$ and a random variable $X : \Omega \to \mathbb{R}$, for simplicity of notation, for every $t \in \mathbb{R}$, we denote $\mathbf{Pr}(X^{-1}(t))$ by $\mathbf{Pr}(X = t)$. We define the *probability mass function (PMF)* of X by

$$p_X(t) = \mathbf{Pr}(X = t).$$

In other words, the PMF is a rule that allocates precise probabilities to distinct values of a random variable.

Now having the PMF of X at hand, the *expected value* or *mean value* of X is defined as

$$E(X) := \sum_{\{t:\mathbf{p}_X(t)>0\}} t \cdot p_X(t),$$

provided that this sum converges absolutely. In this case, we say that the expectation of X is well-defined.

If Ω is finite, then $E(X)$ is well-defined. If Ω is infinite countable, then $E(X)$ exists if

$$E(|X|) := \sum_{\{t:\mathbf{p}_X(t)>0\}} |t| \cdot p_X(t) < \infty,$$

which is the case if $|X| \le M$ for some $M > 0$, that is, $|X(\omega)| \le M$ for all $\omega \in \Omega,$.

Example 1.7. Flipping a coin, let $X : \{H, T\} \to \{0, 1\}$ be given by

$$X(\omega) = \begin{cases} 0 & \text{if } \omega = H, \\ 1 & \text{if } \omega = T. \end{cases}$$

We find the expected value of X. For this sake, we first form the PMF of X:

$$p_X(t) = \begin{cases} \frac{1}{2} & \text{if } t = 0, \\ \frac{1}{2} & \text{if } t = 1, \\ 0 & \text{otherwise.} \end{cases}$$

Thus $E(X) = 0 \cdot \frac{1}{2} + 1 \cdot \frac{1}{2} = \frac{1}{2}$.

Alternatively, we may define the expected value of X as follows:

$$E(X) := \sum_{\omega \in \Omega} p(\omega) \cdot X(\omega).$$

It can be contended that within the context of discrete random variables, the two provided definitions of expected values are equivalent (see Problem 1.7-4). Depending on our requirements and needs, we can utilize either of them as suitable.

In view of the latter definition of the expected value, if we consider each point $\omega \in \Omega$ as a bead of mass $p(\omega)$ concentrated at the point $X(\omega)$ on the real line, then $E(X)$ represents the center of mass for all the beads.

In what follows, we discuss Markov's inequality, a useful tool for using the expected value to control how likely a random variable can be higher than a certain value. For instance, if we have a nonnegative random variable X and the expected value is less than a threshold t, then Markov's inequality tells us that the probability of X being greater than or equal to a is no greater than $\frac{E(X)}{a}$, which is always less than 1. This means that the event $X < a$ has a nonzero probability.

Theorem 1.8 (Markov's inequality). *Let X be a random variable with $R_X \subset [0, \infty)$. Then for any positive real number a, we have*

$$\mathbf{Pr}(X \geq a) \leq \frac{E(X)}{a}.$$

Proof. By definition, $E(X) = \sum_{\{t: p_X(t) > 0\}} t \cdot p_X(t)$. We split this sum into two parts based on whether t is greater than or equal to a, or not. We have

$$E(X) = \sum_{\{t \geq a: \mathbf{p}_X(t) > 0\}} t \cdot p_X(t) + \sum_{\{t < a: \mathbf{p}_X(t) > 0\}} t \cdot p_X(t) \geq \sum_{\{t \geq a: \mathbf{p}_X(t) > 0\}} a \cdot p_X(t) + 0,$$

because in the first sum, we assume that $t \geq a$. Therefore we have

$$E(X) \geq a \sum_{\{t \geq a: \mathbf{p}_X(t) > 0\}} p_X(t) = a\mathbf{Pr}(X \geq a),$$

which completes the proof. ☐

Example 1.9. A coin, biased to land heads with a probability of $\frac{1}{15}$ on each flip, is tossed consecutively 250 times. Applying Markov's inequality, we provide an upper limit on the probability of it landing heads at least 80 times. The biased coin has the probability $p = \frac{1}{15}$ of landing heads on each flip. The expected number of heads in 250 tosses is given by np, where n is the number of tosses, and p is the probability of success: $E(X) = 250 \cdot \frac{1}{15}$. Now we use Markov's inequality with $a = 80$:

$$\mathbf{Pr}(X \geq 80) \leq \frac{\frac{250}{15}}{80} = \frac{5}{24}.$$

Therefore the upper limit on the probability of the biased coin landing heads at least 80 times is $\frac{5}{24}$.

Another answer to the inquiry of "what is the probability that the value of X deviates significantly from its expected value" is provided by Chebyshev's inequality stated as Theorem 1.12. This inequality is applicable to any random variable, not exclusively to nonnegative ones.

Moments. The moments of a random variable are expected values of powers or related functions of the random variable. Precisely speaking, the *kth moment* of X is defined as $\tau_k := E(X^k)$ for every nonnegative integer k. We make the convention that $\tau_0 := E(X^0) = E(1) = 1$. Clearly, the first moment of X is nothing but the expected value of X. If Ω is finite, then the kth moment always exists. If Ω is infinite countable, then k th moment exists if and only if $E(|X|^k) < \infty$. If $E(X^k)$ is not well-defined, then we say that X does not possess the kth moment.

Example 1.10. Let X be a random variable with $R_X = \{1, 2, 3, 4\}$ and the following PMF:

$$p_X(t) = \begin{cases} \frac{1}{8} & \text{if } t = 1, \\ \frac{1}{2} & \text{if } t = 2, \\ \frac{1}{4} & \text{if } t = 3, \\ \frac{1}{8} & \text{if } t = 4, \\ 0 & \text{otherwise.} \end{cases}$$

Then the third moment of X is

$$\tau_3 = E(X^3) = \frac{1}{8} \cdot 1^3 + \frac{1}{2} \cdot 2^3 + \frac{1}{4} \cdot 3^3 + \frac{1}{8} \cdot 4^3 = \frac{151}{8}.$$

Assume that all the moments of X exist. Then under very mild conditions, the moments of X, i. e., the values $\tau_1, \tau_2 \ldots$ determine the CDF F_X of X. For example, this is true if $|X| \le M$ for some $M > 0$. More generally, it is true if there exists $M > 0$ such that

$$\sum_{i=1}^{\infty} \frac{E(X^{2k})}{M^{2k}(2k)!} < \infty.$$

The reason for focusing solely on even moments is as follows: Exploiting the Cauchy–Schwarz inequality (see Appendix A.7.2), we have the inequality

$$|E(X^k)| \le E(|X|^k) = \sum_{\omega \in \Omega} \sqrt{p(\omega)}\left(\sqrt{p(\omega)}|X(\omega)|^k\right)$$

$$\le \sqrt{\sum_{\omega \in \Omega} p(\omega)} \sqrt{\sum_{\omega \in \Omega} p(\omega)X(\omega)^{2k}}$$

$$= 1 \cdot \sqrt{E(X^{2k})}$$

for all k. Hence $\tau_{2k-1}^2 \le \tau_{2(2k-1)}$ for $k \in \mathbb{N}$.

Variance, standard deviation, and Chebyshev's inequality. The *variance* of V is defined as the second moment of $X - E(X)$:

$$\text{Var}(X) := E\left((X - E(X))^2\right) = E(X^2) - E(X)^2.$$

Note that the random variable X takes only one value $E(X)$ if and only if $\text{Var}(X) = 0$. Clearly,

$$E(cX) = cE(X), \quad \text{Var}(cX) = c^2\text{Var}(X) \quad \text{for all } c \in \mathbb{R}.$$

The *standard deviation* of X is defined as $\sigma_X := \sqrt{\text{Var}(X)}$. In general, a larger standard deviation indicates a broader spread of potential values for the random variable.

Example 1.11. When rolling a die, we calculate both the variance and standard deviation. To find the variance, we use the formula

$$\text{Var}(X) = E(X^2) - E(X)^2.$$

First, we find $E(X)$:

$$E(X) = 1 \cdot \frac{1}{6} + 2 \cdot \frac{1}{6} + 3 \cdot \frac{1}{6} + 4 \cdot \frac{1}{6} + 5 \cdot \frac{1}{6} + 6 \cdot \frac{1}{6} = \frac{21}{6}.$$

Next, we find $E(X^2)$:

$$E(X^2) = 1^2 \cdot \frac{1}{6} + 2^2 \cdot \frac{1}{6} + 3^2 \cdot \frac{1}{6} + 4^2 \cdot \frac{1}{6} + 5^2 \cdot \frac{1}{6} + 6^2 \cdot \frac{1}{6} = \frac{91}{6}.$$

Then $\text{Var}(X) = \frac{91}{6} - \left(\frac{21}{6}\right)^2 = \frac{35}{12}$, and $\sigma_X := \sqrt{\text{Var}(X)} = \sqrt{\frac{35}{12}}$.

We have observed that, intuitively, the variance (or, more precisely, the standard deviation) serves as a measure of "spread" or deviation from the mean. Our objective now is to establish a quantitative framework to precisely formalize this intuition. To this end, we use Chebyshev's inequality.

Theorem 1.12 (Chebyshev's inequality). *Let X be a random variable with finite $E(X)$ and $\text{Var}(X)$. Then for all $a > 0$, we have*

$$\mathbf{Pr}(|X - E(X)| \geq a) \leq \frac{\text{Var}(X)}{a^2}.$$

Proof. We start with computing the variance of X:

$$\text{Var}(X) = E\big((X - E(X))^2\big)$$
$$= \sum_{\omega \in \Omega} (X(\omega) - E(X))^2$$
$$= \sum_{\omega \in \Omega, |X(\omega) - E(X)| \geq a} (X(\omega) - E(X))^2 + \sum_{\omega \in \Omega, |X(\omega) - E(X)| < a} (X(\omega) - E(X))^2$$
$$\geq \sum_{\omega \in \Omega, |X(\omega) - E(X)| \geq a} a^2$$
$$= \mathbf{Pr}(|X - E(X)| \geq a) a^2.$$

Dividing both sides by a^2 leads to Chebyshev's inequality. ☐

Historical note. Chebyshev's inequality, proved by the Russian mathematician Pafnuty Chebyshev (Figure 1.5) in 1867, had been previously mentioned by the French statistician Irénée-Jules Bienaymé in 1853. However, there was no accompanying proof for the theory at that time. It was Pafnuty Chebyshev who provided a formal proof for Chebyshev's inequality, and later in 1884, one of his students, Andrey Markov, offered an alternative proof for this result, an important tool in probability theory.

Figure 1.5: Pafnuty Chebyshev (1821–1894). Photo credit: Wikipedia. Source: https://en.wikipedia.org/wiki/Pafnuty_Chebyshev.

Example 1.13. Imagine selecting an individual randomly from a population where the average income is \$60,000 with a standard deviation of \$10,000. Our goal is to determine the probability of selecting an individual with an income of either less than \$40,000 or greater than \$80,000. Given the limited information about the income distribution, we cannot precisely calculate this probability. Nevertheless, we can leverage Chebyshev's inequality to establish an upper bound for it. If the random variable X represents the income, then X falls below \$40,000 or surpasses \$80,000 if and only if $|X - 60000| > 20000$. The probability is

$$\mathbf{Pr}(|X - E(X)| \geq 20000) \leq \frac{\mathrm{Var}(X)}{20000^2} = \frac{100000}{400000} = \frac{1}{4}.$$

1.3 Several random variables

Let us explore the variance of the sum of two random variables X and Y defined on the same probability space. Clearly, $X + Y : \Omega \to \mathbb{R}$ is a random variable such that $(X + Y)(\omega) = X(\omega) + Y(\omega)$. We begin with the linearity of expectation:

$$
\begin{aligned}
E(X + Y) &= \sum_{\omega \in \Omega} p(\omega) \cdot (X + Y)(\omega) \\
&= \sum_{\omega \in \Omega} p(\omega) \cdot (X(\omega) + Y(\omega)) \\
&= \sum_{\omega \in \Omega} p(\omega) \cdot X(\omega) + \sum_{\omega \in \Omega} p(\omega) \cdot Y(\omega) \\
&= E(X) + E(Y).
\end{aligned}
$$

We next inquire about the linearity of the variance:

$$\mathrm{Var}(X + Y) = E\big((X + Y - E(X + Y))^2\big)$$

$$= E((X - E(X))^2 + (Y - E(Y))^2 + 2(X - E(X))(Y - E(Y)))$$
$$\Rightarrow Var(X + Y) = Var(X) + Var(Y) + 2E((X - E(X))(Y - E(Y))). \qquad (1.3.1)$$

This means that variances are additive when $E((X - E(X))(Y - E(Y))) = 0$ but not necessarily in other cases. We will explore this specific case later in this section when we encounter independent random variables.

Joint probability. Given random variables X and Y, the probability $\mathbf{Pr}(X = a, Y = b)$ represents the *joint probability* of X and Y taking specific values a and b, respectively, that is,

$$\mathbf{Pr}(X = a, Y = b) = \mathbf{Pr}(X = a \cap Y = b) = p_{X,Y}(a, b).$$

The function $p_{X,Y}$ above is called the *joint probability mass function* (joint PMF) of X and Y. For random variables X and Y with $R_X = \{a_1, \ldots, a_n\}$ and $R_Y = \{b_1, \ldots, b_m\}$, we form the joint probability table as Table 1.1:

Table 1.1: Joint probability.

X	Y			
	b_1	b_2	$\cdots$	b_m
a_1	$P_{x,y}(a_1, b_1)$	$P_{x,y}(a_1, b_2)$	$\cdots$	$P_{x,y}(a_1, b_m)$
a_2	$P_{x,y}(a_2, b_1)$	$P_{x,y}(a_2, b_2)$	$\cdots$	$P_{x,y}(a_2, b_m)$
$\vdots$	$\vdots$	$\vdots$	$\cdots$	$\vdots$
a_n	$P_{x,y}(a_n, b_1)$	$P_{x,y}(a_n, b_2)$	$\cdots$	$P_{x,y}(a_n, b_m)$

Example 1.14. Roll two dice. Let X be the value on the first die, and let Y be the value on the second die. Both X and Y can assume values between 1 and 6. The joint PMF is given by $p_{X,Y}(i, j) = \frac{1}{36}$ for $1 \le i, j \le 6$. The joint probability table is as Table 1.2:

Table 1.2: Joint probability.

X	Y					
	1	2	3	4	5	6
1	$\frac{1}{36}$	$\frac{1}{36}$	$\frac{1}{36}$	$\frac{1}{36}$	$\frac{1}{36}$	$\frac{1}{36}$
2	$\frac{1}{36}$	$\frac{1}{36}$	$\frac{1}{36}$	$\frac{1}{36}$	$\frac{1}{36}$	$\frac{1}{36}$
3	$\frac{1}{36}$	$\frac{1}{36}$	$\frac{1}{36}$	$\frac{1}{36}$	$\frac{1}{36}$	$\frac{1}{36}$
4	$\frac{1}{36}$	$\frac{1}{36}$	$\frac{1}{36}$	$\frac{1}{36}$	$\frac{1}{36}$	$\frac{1}{36}$
5	$\frac{1}{36}$	$\frac{1}{36}$	$\frac{1}{36}$	$\frac{1}{36}$	$\frac{1}{36}$	$\frac{1}{36}$
6	$\frac{1}{36}$	$\frac{1}{36}$	$\frac{1}{36}$	$\frac{1}{36}$	$\frac{1}{36}$	$\frac{1}{36}$

Random vectors. A *random vector* is a vector whose components are random variables, meaning that the values of the vector are not fixed but are subject to some probability distribution. Each component of the vector represents a different random variable. Formally, if $\mathbf{X} = (X_1, \ldots, X_n)$ where each X_i is a random variable, then $\mathbf{X}$ is a random vector. A random vector has a joint probability distribution, which describes the probabilities of different combinations of values occurring for each component. For instance, consider the random variables X and Y in Example 1.14. Then $\mathbf{X} = (X, Y)$ is a random vector whose each component is a random variable with a discrete uniform distribution from 1 to 6.

Covariance and correlation. When two random variables X and Y are not independent, it is often important to examine the degree of their mutual dependence. This consideration underscores the significance of the concept of covariance given below:

The *covariance* $\mathrm{Cov}(X, Y)$ of X and Y is defined as follows:

$$
\begin{aligned}
\mathrm{Cov}(X, Y) &:= \mathrm{E}((X - \mathrm{E}(X))(Y - \mathrm{E}(Y))) \\
&= \mathrm{E}(XY) - \mathrm{E}(X\mathrm{E}(Y)) - \mathrm{E}(\mathrm{E}(X)Y) + \mathrm{E}(\mathrm{E}(X)\mathrm{E}(Y)) \\
&= \mathrm{E}(XY) - \mathrm{E}(X)\mathrm{E}(Y),
\end{aligned}
\tag{1.3.2}
$$

where for every $\omega \in \Omega$, we define $XY(\omega) = X(\omega)Y(\omega)$ and

$$
\mathrm{E}(XY) = \sum_{a \in R_X} \sum_{b \in R_Y} a \cdot b \cdot \mathbf{Pr}(X = a, Y = b).
$$

Taking the concept of covariance into account, we observe the following:
- The covariance can be viewed as a generalization of variance as the variance of a random variable X is just the covariance of X with itself, $\mathrm{Cov}(X, X) = \mathrm{Var}(X)$.
- the formula for $\mathrm{Var}(X + Y)$ obtained in (1.3.1) may be rewritten as $\mathrm{Var}(X + Y) = \mathrm{Var}(X) + \mathrm{Var}(Y) + 2\mathrm{Cov}(X, Y)$.
- The covariance is symmetric: $\mathrm{Cov}(X, Y) = \mathrm{Cov}(Y, X)$.
- The covariance exhibits linearity in each coordinate, implying two key observations. Initially, constants can be freely factored through either coordinate:

$$
\mathrm{Cov}(aX, Y) = a\,\mathrm{Cov}(X, Y) \quad \text{and} \quad \mathrm{Cov}(X, aY) = a\,\mathrm{Cov}(X, Y) \quad \text{for all } a \in \mathbb{R}.
$$

Second, it preserves sums in each coordinate:

$$
\mathrm{Cov}(X_1 + X_2, Y) = \mathrm{Cov}(X_1, Y) + \mathrm{Cov}(X_2, Y),
$$

and

$$
\mathrm{Cov}(X, Y_1 + Y_2) = \mathrm{Cov}(X, Y_1) + \mathrm{Cov}(X, Y_2).
$$

Both observations are straightforwardly proved, and the proofs are left to the reader interested in delving into the details.

If both variables consistently deviate in the same direction, either both going above their means or below their means simultaneously, then the covariance is positive. Conversely, if they tend to deviate in opposite directions, then the covariance is negative. When X and Y are not strongly related, the covariance is approximately zero.

With the covariance of X and Y at our disposal, we are now prepared to define the correlation of X and Y: For this sake, apply the Cauchy–Schwarz inequality (refer to Appendix A.7.2) to $|E((X - E(X)(Y - E(Y))|$ to derive the inequality

$$\text{Cov}(X, Y)^2 \le \text{Var}(X)\text{Var}(Y).$$

Thus if $\text{Var}(X), \text{Var}(Y) > 0$, then $\frac{\text{Cov}(X,Y)}{\sqrt{\text{Var}(X)}\sqrt{\text{Var}(Y)}} \in [-1, 1]$.

The *correlation coefficient* $\text{Corr}(X, Y)$ of X and Y is defined by

$$\text{Corr}(X, Y) = \frac{\text{Cov}(X, Y)}{\sqrt{\text{Var}(X)}\sqrt{\text{Var}(Y)}} = \frac{\text{Cov}(X, Y)}{\sigma_X \cdot \sigma_Y}.$$

In other words, the correlation coefficient is a normalized version of the covariance.

Furthermore, the *angle* between X and Y is defined as

$$\theta := \arccos(\text{Corr}(X, Y)) \in [0, \pi].$$

In particular, X and Y are called *orthogonal* or *uncorrelated* if $\text{Cov}(X, Y) = 0$.

Example 1.15. Let X be a randomly selected number from the set $\{2, 3, 4, 5\}$, each with an equal likelihood of being chosen. Once X is selected, another random variable Y is drawn from the numbers $\{1, 2, \ldots, X\}$, with each number having an equal probability of being chosen. For instance, if $X = 4$, then Y is drawn from the set $\{1, 2, 3, 4\}$, and the probabilities are given by $\mathbf{Pr}(Y = 1|X = 4) = \frac{1}{4}$, $\mathbf{Pr}(Y = 2|X = 4) = \frac{1}{4}$, $\mathbf{Pr}(Y = 3|X = 4) = \frac{1}{4}$, and $\mathbf{Pr}(Y = 4|X = 4) = \frac{1}{4}$. We first determine the joint PMF of X and Y:

$$p_{X,Y}(a, b) = \mathbf{Pr}(X = a, Y = b) = \mathbf{Pr}(Y = b|X = a) \cdot \mathbf{Pr}(X = a).$$

Clearly, $\mathbf{Pr}(X = a) = \frac{1}{4}$ for every $a \in \{2, 3, 4, 5\}$. Moreover, we have

$$\mathbf{Pr}(Y = b|X = a) = \begin{cases} \frac{1}{a} & \text{if } b \le a, \\ 0 & \text{otherwise.} \end{cases}$$

Therefore, for the joint PMF, we obtain

$$p_{X,Y}(a, b) = \begin{cases} \frac{1}{4a} & \text{if } b \le a, \\ 0 & \text{otherwise.} \end{cases}$$

Thus

$$E(XY) = \sum_a \sum_b a \cdot b \cdot p_{X,Y}(a,b) = \sum_{a=2}^{5} \sum_{b=1}^{a} a \cdot b \cdot \frac{1}{4a} = \sum_{a=2}^{5} \sum_{b=1}^{a} \frac{b}{4}$$

$$= \left(\frac{1}{4} + \frac{2}{4}\right) + \left(\frac{1}{4} + \frac{2}{4} + \frac{3}{4}\right) + \left(\frac{1}{4} + \frac{2}{4} + \frac{3}{4} + \frac{4}{4}\right) + \left(\frac{1}{4} + \frac{2}{4} + \frac{3}{4} + \frac{4}{4} + \frac{5}{4}\right) = \frac{17}{2}.$$

By a straightforward computation we can find $E(X) = \frac{7}{2}$, $E(X^2) = \frac{27}{2}$, $E(Y) = \frac{9}{4}$, and $E(Y^2) = \frac{77}{12}$. Therefore

$$\text{Var}(X) = E(X^2) - E(X)^2 = \frac{27}{2} = \frac{49}{4} = \frac{5}{4},$$

$$\text{Var}(Y) = E(Y^2) - E(Y)^2 = \frac{77}{12} - \frac{81}{16} = \frac{65}{48},$$

$$\text{Cov}(X,Y) = E(XY) - E(X)E(Y) = \frac{17}{2} - \left(\frac{7}{2}\right) \cdot \left(\frac{9}{4}\right) = \frac{5}{8},$$

$$\text{Corr}(X,Y) = \frac{\text{Cov}(X,Y)}{\sigma_X \cdot \sigma_Y} = \frac{\frac{5}{8}}{(\sqrt{\frac{5}{4}}) \cdot (\sqrt{\frac{65}{48}})} = \sqrt{\frac{3}{13}}.$$

Although the variance is often more convenient for computational purposes, its interpretation possesses a challenge due to its expression in squared units. To address this issue, the standard deviation of a random variable is defined as the square root of its variance. A practical, albeit approximate, interpretation suggests that the standard deviation of X roughly indicates how much you would expect the actual value of X to deviate from its expected value $E(X)$.

Likewise, the covariance is frequently "descaled," resulting in the correlation between two random variables. In general, whereas the covariance provides a measure of the joint variability of two variables, the correlation standardizes this measure, facilitating interpretation and comparison across different pairs of variables.

Independent random variables. We previously discussed independent events. Now let us turn our attention to defining the analogous concept for random variables. Let $X, Y : \Omega \rightarrow \mathbb{R}$ be two countable random variables. Then X and Y are called *independent* if

$$\mathbf{Pr}(X = a, Y = b) = \mathbf{Pr}(X = a) \cdot \mathbf{Pr}(X = b) \quad \text{for all } a, b \in \mathbb{R},$$

that is, the *outcome* of the event $X = a$ is independent of the outcome of the event $Y = b$. Random variables which are not independent are called *dependent*.

Example 1.16. Rolling two fair dice involves defining two random variables:
– X, the number obtained on the first die;
– Y, the number obtained on the second die.

Both X and Y are discrete random variables taking values from 1 to 6. Note that the outcome of one die does not influence the outcome of the other; they are independent events. The PMF for each random variable is given by

$$\Pr(X = a) = \Pr(Y = b) = \frac{1}{6} \quad \text{for all } a, b = 1, 2, \dots, 6.$$

On the other hand,

$$\Pr(X = a, Y = b) = \Pr(X = a) \cdot \Pr(Y = b) = \frac{1}{36} \quad \text{for all } a, b = 1, 2, \dots, 6.$$

This suggests that X and Y are independent random variables.

Let X and Y be countable independent random variables. So $X(\Omega) = \{x_i \in \mathbb{R} : i \in \mathcal{I}\}$ and $Y(\Omega) = \{y_i \in \mathbb{R} : i \in \mathcal{I}\}$, where $\mathcal{I}$ is an index set, and Ω is countable. Then

$$E(XY) = \sum_{i,j \in \mathcal{I}} \Pr(X = x_i, Y = y_j) x_i y_j = \sum_{i,j \in \mathcal{I}} \Pr(X = x_i) \Pr(Y = y_j) x_i y_j$$

$$= \left(\sum_{i \in \mathcal{I}} \Pr(X = x_i) x_i \right) \left(\sum_{j \in \mathcal{I}} \Pr(Y = y_j) y_j \right) = E(X)E(Y).$$

Moreover, if X and Y are independent, then $\text{Cov}(X, Y) = 0$ (see Problem 1.7-16), that is, independent random variables are uncorrelated, which implies that

$$\text{Var}(X + Y) = \text{Var}(X) + \text{Var}(Y).$$

In other words, the variance is additive if we deal with independent random variables.

It is worth mentioning that, as discussed earlier, if two random variables X and Y are independent, then they are uncorrelated. However, the converse is not true. For instance, consider X and Y from Example 1.16. We assert that although $X + Y$ and $X - Y$ are uncorrelated, they are not independent;

- $X + Y$ and $X - Y$ are uncorrelated:

$$\text{Cov}(X + Y, X - Y) = \text{Cov}(X, X) + \text{Cov}(X, -Y) + \text{Cov}(Y, X) + \text{Cov}(Y, -Y)$$

$$= \text{Var}(X) - \text{Var}(Y) = 0.$$

- $X + Y$ and $X - Y$ are not independent:

$$\Pr(X + Y = 12, X - Y = 5) = 0,$$

whereas

$$\Pr(X + Y = 12) = \Pr(X = 6) \cdot \Pr(Y = 6) = \frac{1}{36},$$

and

$$\Pr(X - Y = 5) = \Pr(X = 6) \cdot \Pr(Y = 1) = \frac{1}{36},$$

implying

$$\mathbf{Pr}(X + Y = 12, X - Y = 5) \neq \mathbf{Pr}(X + Y = 12) \cdot \mathbf{Pr}(X - Y = 5).$$

Remark 1.17. Having n random variables $X_1, X_2, \ldots, X_n$, we can arrange them into a (row) vector

$$\mathbf{X} = (X_1, X_2, \ldots, X_n).$$

We refer to $\mathbf{X}$ as a *random vector*. The vector $\mathbf{X}$ is n-dimensional because it is composed of n random variables.

Historical note. The covariance and correlation have their roots in the nineteenth century, when early statisticians and mathematicians, notably Francis Galton, began investigating relationships between variables. Galton (Figure 1.6), a pioneering figure in statistics, conducted influential studies in the late 1800s on the correlation between parental heights and their offspring, introducing the concept of regression to the mean. In the late nineteenth and early twentieth centuries, Karl Pearson, another influential statistician, expanded the concept of covariance. In 1896, Pearson not only coined the term "covariance" but also took the lead in formalizing the mathematical representation of the relationship between two variables. Pearson's contributions extended to the introduction of the correlation coefficient, a standardized measure of dependence ranging from –1 to 1. This coefficient provided a clear indication of perfect negative correlation (–1), perfect positive correlation (1), and no correlation (0). A significant milestone was the introduction of matrix notation for covariance and correlation, offering a concise and potent means of simultaneously representing relationships among multiple variables. This matrix approach became the base of multivariate statistics.

Figure 1.6: Francis Galton (1822–1911). Credit: Francis Galton, detail of an oil painting by G. Graef, 1882; in the National Portrait Gallery, London. Source: https://www.britannica.com/biography/Francis-Galton.

Expected value and variance in the context of multiple independent random variables. Given two random variables X and Y over the same probability space, we previously established that

1. $E(X + Y) = E(X) + E(Y)$,
2. $Var(X + Y) = Var(X) + Var(Y) + 2Cov(X, Y)$, and
3. $E(XY) = E(X) \cdot E(Y)$ if X and Y are independent random variables.

We aim to restate these relations for m random variables. To begin, we extend the definition of independence to m random variables: random variables $X_1, \ldots, X_m : \Omega \to \mathbb{R}$ are called *independent* over a countable sample space Ω if

$$\mathbf{Pr}(X_1 = a_1, X_2 = a_2, \ldots, X_m = a_m) = \mathbf{Pr}(X_1 = a_1) \cdot \mathbf{Pr}(X_2 = a_2) \cdots \mathbf{Pr}(X_m = a_m)$$

for all $a_1, a_2, \ldots, a_m \in \mathbb{R}$.

In an easy manner similar to the argument for $m = 2$, we may extend (3) to any finite number of random variables, that is, assuming that $X_i : \Omega \to \mathbb{R}, 1 \le i \le m$, are m random variables, we may easily verify that

$$E(X_1 \cdot X_2 \cdots X_m) = E(X_1) \cdot E(X_2) \cdots E(X_m).$$

Furthermore, (1) and (2) may be extended as follows:

$$E\left(\sum_{i=1}^{m} X_i\right) = \sum_{i=1}^{m} E(X_i), \quad Var\left(\sum_{i=1}^{m} X_i\right) = \sum_{i=1}^{m} Var(X_i) + \sum_{i \ne j} Cov(X_i, X_j). \qquad (1.3.3)$$

The latter relation for the variance may be rewritten for m random variables $X_1, \ldots, X_m$ as follows; since every pair of random variables X_i, X_j are independent for $i \ne j$, we have $Cov(X_i, X_j) = 0$ for all $i \ne j$. Consequently, the variance is additive:

$$Var\left(\sum_{i=1}^{m} X_i\right) = \sum_{i=1}^{m} Var(X_i). \qquad (1.3.4)$$

1.4 Bernoulli, binomial, and Poisson random variables

In various problems, certain classic abstractions of random variables frequently emerge. This section aims to introduce several important discrete distributions. Recognizing that a random variable conforms to one of these structures during problem-solving allows us to employ its precomputed probability mass function, expectation, variance, and other attributes. Such random variables are termed *parametric* because if we can establish that a random variable aligns with one of the parametric types studied, then we only need to specify its parameters. A fitting comparison can be drawn to programming classes, where the process of creating a parametric random variable closely resembles invoking a constructor with specified input parameters.

Bernoulli random variables. A random variable X is referred to as *Bernoulli* when it is defined as $X : \Omega \to \{0,1\}$, which means that X can only take values 0 or 1. If X is a Bernoulli random variable, denoted $X \sim \text{Ber}(p)$, then we have:
-   the probability mass function: $\mathbf{Pr}(X = 1) = p \in [0,1]$ and $\mathbf{Pr}(X = 0) = 1 - p$.
-   $X^k = X$ for all $k \in \mathbb{N}$.
-   the expectation $\mathrm{E}(X) = \mathbf{Pr}(X = 0) \cdot 0 + \mathbf{Pr}(X = 1) \cdot 1 = \mathbf{Pr}(X = 1) = p$.
-   $\mathrm{E}(X^2) = \mathrm{E}(X) = p$.
-   the variance $\text{Var}(X) = \mathrm{E}(X^2) - \mathrm{E}(X)^2 = p(1-p)$.
-   the probability parameter p.

Example 1.18. Rolling a die, our random variable X is assigned the value 1 if the outcome is strictly greater than 2 and 0 otherwise. Consequently, X follows a Bernoulli distribution with probability parameter $\frac{1}{3}$, that is, $X \sim \text{Ber}(\frac{2}{3})$.

With any event $A \subset \Omega$, we associate the following binary (characteristic) random variable $X_A : \Omega \to \mathbb{R}$: $X_A(\omega) = 1$ if and only if $\omega \in A$. Then $\mathrm{E}(X_A) = \mathbf{Pr}(A)$. In other words, characteristic random variables are essentially Bernoulli random variables with parameter p set to the probability of A ($p = \mathbf{Pr}(A)$).

The following fact about Bernoulli random variables is straightforward, and its proof is left to the reader.

Proposition 1.19. *Let Ω be a sample space, and let $X_1, \dots, X_k : \Omega \to \{0,1\}$ be Bernoulli random variables. Let $A_i := \{\omega \in \Omega : X_i(\omega) = 1\}$ for $i = 1, \dots, k$. Then $X_i = X_{A_i}$ and $1 - X_i = X_{A_i^c}$ are Bernoulli for $i = 1, \dots, k$. Furthermore, $X = X_1 \cdot X_2 \cdots X_k$ is Bernoulli, and $X = X_{A_1 \cap A_2 \cap \dots \cap A_k}$. In particular, $\mathrm{E}(X_1 \cdot X_2 \cdots X_k) = \mathbf{Pr}(A_1 \cap A_2 \cap \dots \cap A_k)$.*

Binomial random variables. Random variables $X_1, \dots, X_n : \Omega \to \mathbb{R}$ are called *identically distributed*, abbreviated as i. d., if $F_{X_1} = \dots = F_{X_n}$. Additionally, $X_1, \dots, X_n$ are called *independent identically distributed* random variables, abbreviated as i. i. d., if, in addition to being identically distributed, these variables are also independent.

Let $X_1, \dots, X_n : \Omega \to \{0,1\}$ be Bernoulli. Define $Y = X_1 + \dots + X_n$. Then $Y : \Omega \to \{0,1,\dots,n\}$ have nonnegative integer values in $[0,n]$. The exact distribution of X depends on the *joint* distribution of $X_1, \dots, X_n$. For each integer $k \in [0,n]$, $\mathbf{Pr}(X = k)$ can be expressed as the expectation of the following Bernoulli random variable W_k for $k = 0, 1, \dots, n$. Let $W_0 = (1 - X_1) \cdots (1 - X_n)$. Then $\mathbf{Pr}(Y = 0) = \mathrm{E}(W_0)$. Consider the random variable $U_k = X_1 \dots X_k (1 - X_{k+1}) \cdots (1 - X_n)$. Then U_k is Bernoulli with

$$U_k = 1 \iff X_1 = \dots = X_k = 1, \quad X_{k+1} = \dots = X_n = 0.$$

Hence

$$\mathbf{Pr}(X_1 + \cdots + X_n = k) = E(W_k), \quad W_k = \sum_{1 \le i_1 < \cdots < i_k \le n} X_{i_1} \cdots X_{i_k} \prod_{j \ne i_1, \dots, j \ne i_k} (1 - X_j) \quad (1.4.1)$$

for $k = 0, 1, \dots, n$.

In the case $X_1, \dots, X_n$ are i. i. d. Bernoulli, we can find the distribution of $Y := Y(p, n)$ using one parameter $p = E(X_1) = \cdots = E(X_n)$. Indeed, $Y = k$ if exactly $X_{i_1} = \cdots = X_{i_k} = 1$, $1 \le i_1 < \cdots < i_k \le n$ with probability p^k, while all the other variables take the value 0 with probability $(1-p)^{n-k}$. Hence the probability of the above event is $p^k(1-p)^{n-k}$. Since we can choose $1 \le i_1 < \cdots < i_k \le n$ in $\binom{n}{k}$ ways, it follows that

$$\mathbf{Pr}(Y(n, p) = k) = \binom{n}{k} p^k (1 - p)^{n-k}, \quad k = 0, 1, \dots, n. \quad (1.4.2)$$

The random variable $Y(n, p)$ is called *binomial* with parameters n and p. In other words, to obtain a binomial random variable $Y(n, p)$, we repeat a "Bernoulli experiment", Ber(p), n times independently, and we add up the outcomes. Note that if Y is a binomial random variable with parameters n and p, then we sometimes use the notation $Y \sim \mathrm{Bin}(n, p)$. For $Y \sim \mathrm{Bin}(n, p)$, we have:

- the probability mass function: as we discussed above, $\mathbf{Pr}(Y(n, p) = k) = \binom{n}{k} p^k (1 - p)^{n-k}, k = 0, 1, \dots, n$.
- the expectation $E(Y) = E(X_1 + \cdots + X_n) = E(X_1) + \cdots + \cdots + E(X_n) = np$. This is due to the additivity of expectation and $X_i \sim \mathrm{Ber}(p)$.
- the variance $\mathrm{Var}(Y) = np(1 - p)$. (See Worked-out problem 1.6-2 for a proof.)

Example 1.20. We are tossing a fair coin six times and are interested in counting the total number of heads. The coin flips $X_1, X_2, X_3, X_4, X_5, X_6$ are Ber$(\frac{1}{2})$ random variables, and they are independent by assumption. Consequently, the total number of heads is $Y = X_1 + X_2 + X_3 + X_4 + X_5 + X_6 \sim \mathrm{Bin}(6, \frac{1}{2})$.

Poisson random variables. The Poisson distribution is a probability model employed to represent the likelihood of a specific number of events happening within a fixed time interval, assuming that the events occur independently and at a constant average rate. It is one of the most commonly used distributions in statistics. For example, consider the situation of X representing the number of customers at an ATM in 20-minute intervals.

Rigorously speaking, the countable random variable $X : \Omega \to \mathbb{R}$ is said to be a *Poisson* random variable with parameter a shown as $X \sim \mathrm{Pu}(a)$ if $R_X = \{0, 1, \dots\}$ and its PMF is given by

$$\mathbf{Pr}(\mathrm{Pu}(a) = k) = \begin{cases} e^{-a} \frac{a^k}{k!} & \text{if } k \in R_X, \\ 0 & \text{otherwise.} \end{cases}$$

Note that the parameter a represents the average rate of occurrence of the events. This parameter is also equal to the expected value and variance of the distribution. For $X \sim$

Pu(a), we have:

- the expectation $E(X) = E(Pu(a)) = \sum_{k=0}^{\infty} e^{-a} \frac{a^k}{k!} k = a \sum_{k=1} e^{-a} \frac{a^{k-1}}{(k-1)!} = a$.
- $Var(X) = a$. (See Worked-out problem 1.6-3.)

Example 1.21. On average, 0.53 individuals perished in car accidents weekly in a small California town in 1990. We aim to determine the probability of precisely two fatalities occurring in a specific week during that year, assuming that the weekly death count follows a Poisson distribution:

$$\mathbf{Pr}(X = 2) = e^{-0.53} \frac{(0.53)^2}{2!} \approx 0.082.$$

In the following proposition, we observe that it is possible to derive Pu(a) as the limit of the binomial with specific parameters.

Proposition 1.22. *Let $Y(n, p_n) = X_{1,n} + \cdots + X_{n,n}$, where $X_{1,n}, \ldots, X_{n,n}$ are i. i. d. Bernoulli with $E(X_{1,n}) = p_n \in [0,1]$ for $n \in \mathbb{N}$. Assume that there exists a subsequence $1 \le n_1 < n_2 < \cdots$ such that $\lim_{m \to \infty} E(Y(n_m, p_{n_m})) = \lim_{m \to \infty} n_m p_{n_m} = a$. Then $X_{n_1}, X_{n_2}, \ldots$ converge in probability to Pu(a), that is,*

$$\lim_{m \to \infty} \mathbf{Pr}\big(Y(n_m, p_{n_m}) = k\big) = e^{-a} \frac{a^k}{k!} \quad \text{for } k = 0, 1, \ldots.$$

Proof. Note that

$$\mathbf{Pr}(Y(n, p) = k) = \binom{n}{k} p^k (1-p)^{n-k} = \frac{1}{k!}\left(1 - \frac{1}{n}\right) \cdots \left(1 - \frac{k-1}{n}\right)(np)^k (1-p)^n (1-p)^{-k}.$$

Now

$$\lim_{m \to \infty} (n_m p_{n_m})^k = a^k, \quad \lim_{m \to \infty} (1 - p_{n_m})^{n_m} = e^{-a}, \quad \lim_{m \to \infty} (1 - p_{n_m})^{-k} = 1,$$

and the proposition follows. ☐

It is possible to deduce the conclusion of this proposition assuming much less than i. i. d. Bernoulli $X_{1,n}, \ldots, X_{n,n}$.

Theorem 1.23. *Let $1 \le n_1 < n_2 < \cdots$ be an increasing sequence of integers. Let $Z_{1,m}, \ldots, Z_{n_m,m}$ be Bernoulli. Let $Y_m := \sum_{i=1}^{n_m} Z_{i,m}$. Suppose that for each k,*

$$\lim_{m \to \infty} E\left(\sum_{1 \le i_1 < \cdots < i_k \le n_m} Z_{i_1,m} \cdots Z_{i_k,m}\right) = \frac{a^k}{k!} \quad \text{for } k = 0, 1, \ldots \text{ and } a \ge 0. \tag{1.4.3}$$

Then $Y_1, Y_2, \ldots$ converges in probability to Pu(a), that is,

$$\lim_{m \to \infty} \mathbf{Pr}(Y_m = k) = e^{-a} \frac{a^k}{k!} \quad \text{for } k = 0, 1, \ldots. \tag{1.4.4}$$

Proof. We first prove (1.4.4) for $k = 0$. Note that

$$Y_m = 0 \iff Z_{1,m} = \cdots = Z_{n_m,m} = 0 \iff \prod_{i=1}^{n_m}(1 - Z_{i,m}) = 1.$$

Hence $\mathbf{Pr}(Y_m = 0) = \mathrm{E}(\prod_{i=1}^{n_m}(1 - Z_{i,m}))$. Use the arguments of the proof of (1.5.6) to deduce that

$$1 + \sum_{k=1}^{2q-1}(-1)^k \sum_{1 \le i_1 < \cdots < i_k \le n_m} \mathrm{E}(Z_{i_1,m} \cdots Z_{i_k,m}) \le \mathrm{E}\left(\prod_{i=1}^{n_m}(1 - Z_{i,m})\right) = \mathbf{Pr}(Y_m = 0)$$

$$\le 1 + \sum_{k=1}^{2p}(-1)^k \sum_{1 \le i_1 < \cdots < i_k \le n_m} \mathrm{E}(Z_{i_1,m} \cdots Z_{i_k,m}).$$

Let $m \to \infty$ and use assumption (1.4.3) to deduce

$$1 + \sum_{k=1}^{2q-1}(-1)^k \frac{a^k}{k!} \le \liminf_{m\to\infty} \mathrm{E}(Y_m) \le \limsup_{m\to\infty} \mathrm{E}(Y_m) \le 1 + \sum_{k=1}^{2p}(-1)^k \frac{a^k}{k!}.$$

Note that $e^{-a} = \lim_{l\to\infty} 1 + \sum_{k=1}^{l}(-1)^k \frac{a^k}{k!}$. Let $p, q \to \infty$ in the above inequalities to deduce (1.4.4) for $k = 0$.

To prove (1.4.4), we need to use (1.4.1). Let

$$W_{k,m} = \sum_{1 \le j_1 < \cdots < j_k \le n_m} Z_{j_1,m} \cdots Z_{j_k,m} \prod_{j \ne j_1, \ldots, j \ne j_k} (1 - Z_{j,m}).$$

For each $\prod_{j \ne j_1, \ldots, j \ne j_k} (1 - Z_{j,m})$, use inequalities (1.5.2) with fixed p and q. This gives lower and upper bounds for $\mathrm{E}(W_{k,m})$. Let $m \to \infty$ and use (1.4.3) to obtain lower and upper bounds on $\liminf_{m\to\infty} \mathrm{E}(W_{k,m}) \le \limsup_{m\to\infty} \mathrm{E}(W_{k,m})$ as in the case $k = 0$. Now let $p, q \to \infty$ to deduce (1.4.4) for all $k \ge 1$. $\square$

Remark 1.24. We note that the assumptions of Proposition 1.22 imply the conditions of Theorem 1.23. Indeed, $\mathrm{E}(X_{i_1,n} \cdots X_{i_k,n}) = p_n^k$ for all $1 \le i_1 < \cdots < i_k \le n$. Hence $\mathrm{E}(\sum_{1 \le i_1 < \cdots < i_k \le n} X_{i_1,n} \cdots X_{i_k,n}) = \binom{n}{k}p_n^k$. Let $Z_{i,m} = X_{i,n_m}$ for $i = 1, \ldots, n_m$. The assumption that $\lim_{m\to\infty} n_m p_{n_m} = a$ yields (1.4.3) for $k \in \mathbb{Z}_+$.

Historical note. Tracing the roots of these distributions, James Bernoulli (Figure 1.7) (1654–1705), a Swiss mathematician, made a groundbreaking contribution with his work "Ars Conjectandi" in 1713. This represented a pivotal moment in the development of probability theory, as it consolidated counting concepts and provided a proof for the binomial theorem. Siméon-Denis Poisson (Figure 1.8) (1781–1840), a mathematics professor at the Faculté des Sciences, introduced the Poisson distribution in his 1837 text "Recherchés sur la probabilité des jugements en matiére criminelle et en matiére civile."

Figure 1.7: James Bernolli (1654–1705). Credit: Portrait of Jacob Bernoulli (1686, around 32 years old), painted by his brother Nikolaus (1662–1716). Source: https://link.springer.com/article/10.1007/s00283-021-10072-y/figures/1.

Figure 1.8: Siméon-Denis Poisson (1781–1840). Credit: Siméon-Denis Poisson, detail of a lithograph by François-Séraphin Delpech after a portrait by N. Maurin. Source: https://www.britannica.com/biography/Simeon-Denis-Poisson.

Importantly, these scholars aimed to construct intricate abstract models for real-world phenomena, contributing to the evolution of modeling tools such as calculus.

1.5 Inclusion–exclusion principle

In this section, we establish the probabilistic form of the inclusion–exclusion principle and its consequences. It is worth noting that this principle takes on various formulations, some of which are specifically tailored based on counting techniques in combinatorics.

Let us commence with the version designed for two events:

Proposition 1.25. *For any events A and B in $\mathcal{F}$, we have*

$$\mathbf{Pr}(A \cup B) = \mathbf{Pr}(A) + \mathbf{Pr}(B) - \mathbf{Pr}(A \cap B).$$

Proof. We make use of the simple observation that A and $B \setminus A$ are exclusive events and their union is $A \cup B$:

$$\mathbf{Pr}(A \cup B) = \mathbf{Pr}\big(A \cup (B \setminus A)\big) = \mathbf{Pr}(A) + \mathbf{Pr}(B \setminus A).$$

On the other hand, $B \setminus A$ and $B \cap A$ represent mutually exclusive events, and their union is equal to B. Thus

$$\mathbf{Pr}(B) = \mathbf{Pr}\big((B \setminus A) \cup (B \cap A)\big) = \mathbf{Pr}(B \setminus A) + \mathbf{Pr}(B \cap A).$$

The proof of the statement is derived from the difference of the two above relations. □

The inclusion–exclusion principle has the following consequence.

Proposition 1.26. *For any events A and B in $\mathcal{F}$, we have*

$$\mathbf{Pr}(A \cap B) \geq \mathbf{Pr}(A) + \mathbf{Pr}(B) - 1.$$

Proof. It is immediate from the inclusion–exclusion principle:

$$\mathbf{Pr}(A \cap B) = \mathbf{Pr}(A) + \mathbf{Pr}(B) - \mathbf{Pr}(A \cup B) \geq \mathbf{Pr}(A) + \mathbf{Pr}(B) - 1.$$ □

Next, the generalized versions of Propositions 1.25 and 1.26 apply to n events. To derive these versions, we require the following two lemmas.

Lemma 1.27. *Let $x_1, \ldots, x_n \in \mathbb{C}$. Then*

$$(1 - x_1) \cdot (1 - x_2) \cdots (1 - x_n) = 1 + \sum_{k=1}^{n} (-1)^k \sum_{1 \leq i_1 < \cdots < i_k \leq n} x_{i_1} \cdots x_{i_k}. \tag{1.5.1}$$

Assume furthermore that $x_1, \ldots, x_n \in \{0, 1\}$. Then for any even integer $2p \in [0, n]$ and odd integer $2q - 1 \in [1, n]$, we have the inequalities

$$1 + \sum_{k=1}^{2q-1} (-1)^k \sum_{1 \leq i_1 < \cdots < i_k \leq n} x_{i_1} \cdots x_{i_k} \leq \prod_{i=1}^{n} (1 - x_i) \leq 1 + \sum_{k=1}^{2p} (-1)^k \sum_{1 \leq i_1 < \cdots < i_k \leq n} x_{i_1} \cdots x_{i_k}. \tag{1.5.2}$$

Proof. Equality (1.5.1) is straightforward and can be proven by induction on n. Inequalities (1.5.2) are proved as follows. Assume that m out of n variables $x_1, \ldots, x_n$ are equal to 1. If $m = 0$, then $x_1 = \cdots = x_n = 0$, and we have that all the expressions in (1.5.2) are equal to 1. Hence (1.5.2) holds. Assume that $m \in [1, n]$. Without loss of generality, we may assume that $x_1 = \cdots = x_m = 1$ and $x_{m+1} = \cdots x_n = 0$. In that case, $(1 - x_1) \cdots (1 - x_n) = 0$. Observe next that $x_{i_1} \ldots x_{i_l} = 0$ for all $1 \leq i_1 < \cdots < i_l \leq n$ and $l > m$. It follows that

$$\sum_{1 \leq i_1 < \cdots i_l \leq n} x_{i_1} \cdots x_{i_l} = \binom{m}{l} \quad \text{for all integers } l \in [1, n].$$

Indeed, this equality corresponds to choosing l elements $x_{i_1}, \ldots, x_{i_l}$ out of $\{x_1, \ldots, x_m\}$ that are all equal to 1. Thus (1.5.2) is equivalent to

$$\sum_{k=0}^{2q-1} (-1)^k \binom{m}{k} \leq 0 \leq \sum_{k=0}^{2p} (-1)^k \binom{m}{k} \tag{1.5.3}$$

for all $p \in \mathbb{Z}_+$ and $q \in \mathbb{N}$. Notice that the sequence $\binom{m}{l}$ is nondecreasing for $l = 0, 1, \ldots, \lceil \frac{m}{2} \rceil$. Since

$$\sum_{k=0}^{2q-1} (-1)^k \binom{m}{k} = \sum_{k=0}^{q-1} \binom{m}{2k} - \binom{m}{2k+1},$$

it follows that for $2q - 1 \leq \lceil \frac{m}{2} \rceil$, the first inequality in (1.5.3) holds. Since $\binom{m}{0} = 1$, we clearly have the second inequality in (1.5.3) for $p = 0$. For $p \geq 1$, we have the identity

$$\sum_{k=0}^{2p} (-1)^k \binom{m}{k} = 1 + \sum_{k=1}^{2p} \binom{m}{2k} - \binom{m}{2k-1}.$$

Hence for $2p \leq \lceil \frac{m}{2} \rceil$, we deduce the second inequality in (1.5.3).

As $\binom{m}{l} = 0$ for $l > m$, it suffices to prove (1.5.3) for $m \geq 2q - 1, 2p \geq \lceil \frac{m}{2} \rceil$. Notice that $0 = (1-1)^m = \sum_{k=0}^{m} (-1)^k \binom{m}{k}$. Subtract this identity from both sides of (1.5.3) and use the identities $\binom{m}{k} = \binom{m}{m-k}$ for $k = 1, \ldots, m$ to deduce the cases $m \geq 2q - 1, 2p \geq \lceil \frac{m}{2} \rceil$ from the cases $2q - 1, 2p \leq \lceil \frac{m}{2} \rceil$. □

The subsequent straightforward lemma serves as a component in establishing the proof of our main theorem in this section. Given two random variables $X, Y : \Omega \to \mathbb{R}$, we say that X is *less than or equal* to Y and denote this by $X \leq Y$ if $X(\omega) \leq Y(\omega)$ for all $\omega \in \Omega$.

Lemma 1.28. *Given two random variables $X, Y : \Omega \to \mathbb{R}$, if $X \leq Y$, then $\mathrm{E}(X) \leq \mathrm{E}(Y)$.*

Proof.

$$X \leq Y \Rightarrow \mathrm{E}(X) = \sum_{\omega \in \Omega} p(\omega) X(\omega) \leq \sum_{\omega \in \Omega} p(\omega) Y(\omega) = \mathrm{E}(Y). \tag{1.5.4}$$

□

We are now ready to state and prove the main theorem of this section.

Theorem 1.29. *Let $A_1, \ldots, A_n \subset \Omega$ be n events in a sample space Ω. Then*

$$\mathbf{Pr}(A_1 \cup A_2 \cup \cdots \cup A_n) = \sum_{i=1}^{n} \mathbf{Pr}(A_i) + \sum_{k=2}^{n} (-1)^{k-1} \sum_{1 \leq i_1 < \cdots < i_k \leq n} \mathbf{Pr}(A_{i_1} \cap \cdots \cap A_{i_k}). \tag{1.5.5}$$

Furthermore, for any even integer $2p \in [1, n]$ and odd integer $2q - 1 \in [1, n]$, we have

$$\sum_{k=1}^{2p} (-1)^{k-1} \sum_{1 \leq i_1 < \cdots < i_k \leq n} \mathbf{Pr}(A_{i_1} \cap \cdots \cap A_{i_k}) \leq \mathbf{Pr}(\cup_{i=1}^{n} A_i)$$

$$\le \sum_{k=1}^{2q-1} (-1)^{k-1} \sum_{1\le i_1 < \cdots < i_k \le n} \mathbf{Pr}(A_{i_1} \cap \cdots \cap A_{i_k}). \tag{1.5.6}$$

Remark 1.30. Intuitively, when we sum the probabilities, we end up double-counting all the two-intersections. To rectify this, we subtract the second sum, taking into consideration that each two-intersection is present exactly once in the set $\{A_{i_1} \cap A_{i_2} : 1 \le i_1 < i_2 \le n\}$. However, this approach leads to a complication: we have now accounted for all three-intersections three times, subsequently subtracted them three times, necessitating the addition of these intersections once to avoid further complications. This process then poses challenges when dealing with four-intersections and beyond.

Proof of Theorem 1.29. Let $X_i = X_{A_i}$, $Y_i = X_{A_i^c} = 1 - X_i$, $i = 1, \ldots, n$. Then

$$\mathbf{Pr}(\cup_{i=1}^n A_i) = 1 - \mathbf{Pr}((\cup_{i=1}^n A_i)^c) = 1 - \mathbf{Pr}(\cap_{i=1}^n A_i^c)$$
$$= 1 - \mathrm{E}(Y_1 \cdots Y_n) = 1 - \mathrm{E}((1 - X_1) \cdots (1 - X_n)).$$

Use expansion (1.5.1) and Proposition 1.19 to deduce

$$\mathrm{E}((1 - X_1) \cdots (1 - X_n)) = \mathrm{E}\left(1 + \sum_{k=1}^n (-1)^k \sum_{1\le i_1 < \cdots < i_k \le n} X_{i_1} \cdots X_{i_k} \right)$$

$$= 1 + \sum_{k=1}^n (-1)^k \sum_{1\le i_1 < \cdots < i_k \le n} \mathrm{E}(X_{i_1} \cdots X_{i_k})$$

$$= 1 + \sum_{k=1}^n (-1)^k \sum_{1\le i_1 < \cdots < i_k \le n} \mathbf{Pr}(A_{i_1} \cap \cdots \cap A_{i_k}).$$

Combine the above two equalities to deduce (1.5.5). Since each $X_i \in \{0,1\}$, we can apply inequality (1.5.2) to deduce

$$1 + \sum_{k=1}^{2q-1} (-1)^k \sum_{1\le i_1 < \cdots < i_k \le n} X_{i_1} \cdots X_{i_k} \le \prod_{i=1}^n (1 - X_i) \le 1 + \sum_{k=1}^{2p} (-1)^k \sum_{1\le i_1 < \cdots < i_k \le n} X_{i_1} \cdots X_{i_k}.$$

Take the expected value of all the three random variables appearing in the above inequality, then use Lemma 1.28 and the above arguments to obtain (1.5.6). $\quad\square$

Remark 1.31. Equality (1.5.5) is called the *inclusion–exclusion principle*. Inequalities (1.5.6) are called the *Bonferroni* inequalities.

Historical note. The inclusion–exclusion principle, a combinatorial counting method extensively applied in probability theory and combinatorics, has historical roots in the contributions of mathematicians such as Euler and Möbius. Euler, especially, delved into the idea of enumerating elements in sets featuring overlapping components. The contemporary articulation of the inclusion–exclusion principle is commonly credited

Figure 1.9: James Joseph Sylvester (1814–1897). Credit: Wikipedia. Source: https://en.wikipedia.org/wiki/James_Joseph_Sylvester.

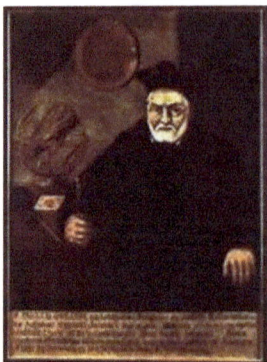

Figure 1.10: Pual Guldin (1577–1643). Credit: Wikipedia. Source: https://en.wikipedia.org/wiki/Paul_Guldin.

to mathematicians such as J. J. Sylvester (Figure 1.9) and Paul Guldin (Figure 1.10). Sylvester, during the nineteenth century, played a pivotal role in advancing combinatorics, thereby establishing the groundwork for the explicit formulation of this principle.

1.6 Worked-out problems

1. Show that Markov's inequality is tight.

   Solution. We find a probability distribution for a random variable X with $\mathbf{Pr}(X \geq a) = \frac{\mathrm{E}(X)}{a}$. Let X be a random variable with $R_X = \{0, 4\}$ and $\mathbf{Pr}(X = 0) = \frac{15}{16}$, $\mathbf{Pr}(X = 4) = \frac{1}{16}$. Then $\mathrm{E}(X) = 4 \cdot \frac{1}{14} = \frac{1}{4}$. On the other hand, invoking Markov's inequality for $a = 4$, we have

$$\mathbf{Pr}(X \geq 4) \leq \frac{E(X)}{4} = \frac{1}{16}.$$

This is exactly $\mathbf{Pr}(X \geq 4) = \mathbf{Pr}(X = 4) = \frac{1}{16}$.

2. Given a binomial random variable $Y \sim Y(n, p)$ with parameters n and p, show that

$$\mathrm{Var}(Y) = np(1 - p).$$

Solution. As $Y(n, p) = \sum_{i=1}^{n} Y_i$, where $Y_1, \ldots, Y_n$ are independent equidistributed random variables, we obtain

$$\mathrm{Var}(Y) = \sum_{i=1}^{n} \mathrm{Var}(Y_i) = np(1 - p).$$

3. Given a Poisson random variable $X \sim \mathrm{Pu}(a)$, prove that $\mathrm{Var}(X) = a$.

Solution. First, we compute

$$
\begin{aligned}
E(X^2) &= \sum_{k \geq 0} k^2 \mathbf{Pr}(X = k) \\
&= \sum_{k \geq 0} k^2 e^{-a} \frac{a^k}{k!} \\
&= ae^{-a}\left(\sum_{k \geq 1} (k - 1)\frac{a^{k-1}}{(k-1)!} + \sum_{k \geq 1} \frac{a^{k-1}}{(k-1)!} \right) \\
&= ae^{-a}\left(a\sum_{k \geq 2} \frac{a^{k-2}}{(k-2)!} + \sum_{k \geq 1} \frac{a^{k-1}}{(k-1)!} \right) \\
&= ae^{-a}\left(a\sum_{i \geq 0} \frac{a^i}{i!} + \sum_{j \geq 0} \frac{a^j}{j!} \right) \\
&= ae^{-a}(ae^a + e^a) \\
&= a^2 + a.
\end{aligned}
$$

The latter relation arises from the Taylor series expansion of the exponential function: $e^x = \sum_{n=0}^{\infty} \frac{e^x}{n!}$ for every $x \in \mathbb{R}$. On the other hand, as we discussed previously, $E(X) = a$. Therefore $\mathrm{Var}(X) = E(X^2) - E(X)^2 = a^2 + a - a^2 = a$.

4. Let $X_1, X_2, \ldots$ be i. i. d. random variables with finite mean μ and finite variance σ^2. Let M_n denote the average of the first n X_is, i. e.,

$$M_n = \frac{1}{n}(X_1 + X_2 + \cdots + X_n).$$

Then, for any $\epsilon > 0$, we have

$$\lim_{n \to \infty} \mathbf{Pr}(|M_n - \mu| \geq \epsilon) = 0.$$

Solution. By the linearity of expectation we have

$$E(M_n) = \frac{1}{n}(E(X_1) + \cdots + E(X_n)) = \frac{n\mu}{n} = \mu.$$

To find the variance of M_n, note that

$$\text{Var}(M_n) = \text{Var}\left(\frac{1}{n}(X_1 + X_2 + \cdots + X_n)\right).$$

Since $X_1, X_2, \ldots, X_n$ are i. i. d., we have

$$\text{Var}(M_n) = \frac{1}{n^2}\text{Var}(X_1 + X_2 + \cdots + X_n).$$

Using the fact that the variance of the sum of i. i. d. random variables is $n\sigma^2$, we get

$$\text{Var}(X_1 + X_2 + \cdots + X_n) = n\sigma^2,$$

and thus

$$\text{Var}(M_n) = \frac{1}{n^2} \cdot n\sigma^2 = \frac{\sigma^2}{n}.$$

Applying Chebyshev's inequality, we obtain

$$\mathbf{Pr}(|M_n - \mu| \geq \epsilon) \leq \frac{\text{Var}(M_n)}{\epsilon^2} = \frac{\sigma^2/n}{\epsilon^2} = \frac{\sigma^2}{n\epsilon^2}.$$

As $n \to \infty$, $\frac{\sigma^2}{n\epsilon^2} \to 0$. Hence

$$\lim_{n\to\infty} \mathbf{Pr}(|M_n - \mu| \geq \epsilon) = 0.$$

Notice that the result is called *the law of large numbers*.

1.7 Problems

1. Let $(\Omega, \mathcal{F}, \mathbf{Pr})$ be a probability space, and let $A, B \in \mathcal{F}$. Prove that:
   (a) $A \cap B \in \mathcal{F}$.
   (b) $A \setminus B \in \mathcal{F}$.
   (c) $A \triangle B = (A \setminus B) \cup (B \setminus A) \in \mathcal{F}$.
   (d) if Ω is a finite set, then $|\mathcal{F}|$ is an even number.
   (e) A and B are independent if and only if A and $\Omega \setminus B$ are independent.
   (f) if $\mathbf{Pr}(A), \mathbf{Pr}(B) > 0$, and $\mathbf{Pr}(A|B) = \mathbf{Pr}(A)$, then $\mathbf{Pr}(B|A) = \mathbf{Pr}(B)$.
2. Prove that if $\mathcal{F}$ is the power set of Ω, then all functions that map Ω into a countable subset of $\mathbb{R}$ are discrete random variables.

3. Let $\mathcal{A} = \{A_n\}$ be a sequence of events, each depending on n. We say that $\mathcal{A}$ *occurs with high probability* (often shortened to w. h. p.) if $\lim_{n\to\infty} \mathbf{Pr}(A_n) = 1$. Now let $\mathcal{A} = \{A_n\}$ be the event that the fraction of heads in n (fair) coin tosses falls within the range $(0.4, 0.6)$. Show that $\mathcal{A}$ occurs with high probability.

4. Given a discrete random variable $X : \Omega \to \mathbb{R}$, prove the following alternative formula for expected value of X:

$$E(X) = \sum_{\omega\in\Omega} p(\omega) \cdot X(\omega).$$

5. Prove the Cauchy–Schwarz inequality for expectations: Given two random variables X and Y, prove that $E(XY)^2 \le E(X^2)E(Y^2)$ and that the equality holds if and only if $X = cY$ for some constant $c \in \mathbb{R}$.

6. The pigeonhole principle asserts that if m balls are distributed randomly into n bins, then there must be at least one bin containing $\lceil \frac{m}{n} \rceil$ balls. How is the Markov inequality equivalent to the pigeonhole principle?

7. Let X be a random variable, and let $f : \mathbb{R} \to \mathbb{R}$ be an increasing function.
   (a) Prove that $\mathbf{Pr}(X \ge a) = \mathbf{Pr}(f(X) \ge f(a))$.
   (b) Combine (a) and Markov's inequality to prove Chebyshev's inequality.

8. A biased coin has a probability of $\frac{1}{5}$ landing heads. The coin is flipped 180 times. Apply Markov's inequality to establish an upper limit on the probability of obtaining heads at least 120 times. Improve this bound by employing Chebyshev's inequality.

9. Show that Chebyshev's inequality is tight (see Worked-out problem 1.6-1.)

10. For any constant $a, b \in \mathbb{R}$ and random variable X, prove that $Var(aX + b) = a^2 Var(X)$.

11. Prove Proposition 1.19.

12. (a) Prove that $Var(X) \ge 0$ for every random variable X.
    (b) Use part (a) to investigate whether there is a random variable X for which $E(X) = 5$ and $E(X^2) = 12$.

13. Let X is a random variable with $E(X^2) = 0$.
    (a) Prove that $\mathbf{Pr}(X = 0) = 1$.
    (b) Consequently, deduce that if $Var(X) = 0$, then $\mathbf{Pr}(X = E(X)) = 1$, where $E(X)$ is assumed to be finite.

14. For any constant $a, b, c, d \in \mathbb{R}$ and random variables X and Y, prove that

$$Cov(aX + c, bY + d) = ab\, Cov(X, Y).$$

15. Prove that $Corr(aX + b, Y) = Corr(X, Y)$ for all $a > 0$ and $b \in \mathbb{R}$.

16. Prove that if X and Y are independent random variables, then $Cov(X, Y) = 0$.

17. Let X and Y be two independent random variables. Let $f, g : \mathbb{R} \to \mathbb{R}$. Prove that:
    (a) $f(X)$ and $g(Y)$ are random variables.
    (b) $f(X)$ and $g(Y)$ are independent.

18. Consider the random variable X as the number of heads obtained after flipping a coin four times. It follows a binomial distribution with parameters $n = 4$ and $p = \frac{1}{2}$. Find the probability associated with each distinct value of R_X.

19. Let X and Y be two random variables.
    (a) For $X \sim \text{Bin}(n, p)$ and $Y \sim \text{Bin}(m, p)$, show that $X + Y \sim \text{Bin}(m + n, p)$.
    (b) For $X \sim \text{Pu}(a)$ and $Y \sim \text{Pu}(b)$, show that $X + Y \sim \text{Pu}(a + b)$.

20. Let $A_1, \ldots, A_n$ be n events on the same probability space. Use induction on n to prove that $\mathbf{Pr}(\bigcup_{i=1}^{n}(A_i)) \leq \sum_{i=1}^{n} \mathbf{Pr}(A_i)$. (This inequality is known as *Boole's inequality*, asserting that the probability of at least one event occurring is not greater than the sum of the probabilities of the individual events in the collection. It is evident that Boole's inequality is specific of the Bonferroni inequalities, appeared in Theorem 1.29.)

21. A circular table has 16 seats, numbered in a clockwise direction. The guests attending dinner comprise 8 king/queen pairs. The queens take random seats in the odd-numbered positions, and the kings are seated randomly between them. Let X represent the count of queens sitting adjacent to their corresponding kings. Determine $E(X)$ and $\text{Var}(X)$.

2 Graphs

A *directed graph* is composed of edges with designated directions, indicating connections from one vertex to another by arrows. On the other hand, an *undirected graph* lacks arrowed edges, representing bidirectional connections between nodes. Although similar to a directed graph, it allows for connections that can be traversed in both directions. Consider, for instance, a scenario where vertices symbolize individuals at a gathering, and an edge exists between two individuals if they shake hands. In this context the graph is undirected because any person A can shake hands with person B only if B reciprocates the gesture. On the contrary, if an edge from person A to person B indicates that A owes money to B, the graph is directed. This directionality arises as owing money does not inherently imply reciprocal indebtedness.

Outline. This chapter is divided into two sections. The first part focuses on undirected graphs, exploring related concepts and results, whereas the second section is devoted to directed graphs. In the undirected graph setting, we begin by exploring fundamental concepts such as the definition of a graph, its related notions, and the degree-sum theorem. We then study connectivity in graphs, including trees, forests, and bipartite graphs. Classification theorems for trees and bipartite graphs are also introduced. Moving forward, we examine Eulerian circuits and trails, classifying all such graphs based on degree criteria. Transitioning to the directed graph setting, we introduce analogous concepts for directed graphs. Topics include directed graph connectivity, acyclic directed graphs, and a decomposition theorem for acyclic digraphs. We finally delve into a specific type of connectivity in digraphs known as *strongly connected digraphs*. The interested reader is referred to [6, 7, 45] for a comprehensive study of graph theory.

Concerning proofs, note that many statements in graph theory can be proved by using the *principle of induction*. In Appendix A.3, we elucidate the preferred form of induction we will frequently employ, known as the *strong principle of induction*.

2.1 Undirected graphs

Warm up. An *undirected* graph is a pair $G := (V, E)$ consisting of:
1. V, a finite set of *vertices*,
2. E, a finite set of *edges*, which are unordered pairs of distinct vertices,

$$E \subset \{(a,b) : a, b \in V\}.$$

In other words, each edge is an unordered pair of $(a,b)(= ab)$ of two distinct vertices $a \neq b \in V$. Sometimes, we let $V = V(G)$ and $E = E(G)$ to emphasize that V and E correspond to the graph G. In this section, all the graphs are assumed to be undirected.

https://doi.org/10.1515/9783111337388-002

A *loop* is an edge that starts and finishes in the same node. *Multiple edges* are two or more edges connecting the same two vertices. A *simple graph* is a graph that does not have more than one edge between any two vertices and no edge starts and ends at the same vertex. In other words, a simple graph is a graph without loops and multiple edges. Throughout this book, unless explicitly mentioned otherwise, all graphs are assumed to be simple.

The *cardinality* $|V|$ is called the *order* of G. Let $n = |V|$. Then it is convenient to identify $V = [n] := \{1, \dots, n\}$. The *cardinality* $|E|$ is called the *size* of G. A graph G is called *trivial* if it contains only one vertex and no edge.

Two graphs $G = (V, E)$ and $H = (W, F)$ of the same order and size are called *isomorphic* if there exists a *bijection* $\phi : V \to W$ such that $(u, v) \in E$ if and only if $(\phi(u), \phi(v)) \in F$. We write $G \cong H$ for isomorphic graphs G and H. A *subgraph* S of G is a graph whose vertex set and edge set are subsets of those in G. We highlight the following families of subgraphs:

- *Spanning subgraph*: A spanning subgraph is a subgraph that contains all the vertices of the original graph, that is, a spanning subgraph of a graph is a subgraph obtained by edge deletions only while keeping all vertices.
- *Vertex-induced subgraph*: A subgraph obtained from the graph G solely by deleting vertices is called a vertex-induced subgraph (or induced subgraph) of G. If X represents the set of deleted vertices, then the induced subgraph is symbolized as $G - X$. When $Y = V(G) \setminus X$, the induced subgraph is denoted $G[Y]$ and is referred to as the subgraph of G induced by the vertex set Y.
- *Edge-induced subgraph*: Let S be a set of edges in a graph G. The edge-induced subgraph $G[S]$ is a subgraph of G defined by the edges in S, with its vertex set comprising all the endpoints of the edges in S.

Example 2.1. Consider the following graph G:

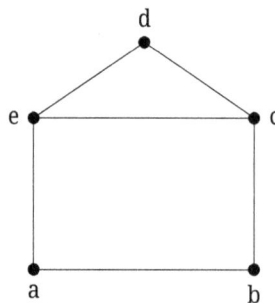

Then, in the following table, we list various subgraphs of G:

A spanning subgraph

$G[Y]$, where $Y = \{a, b, c, d\}$

$G[S]$, where $S = \{ab, bc, ce, ea\}$

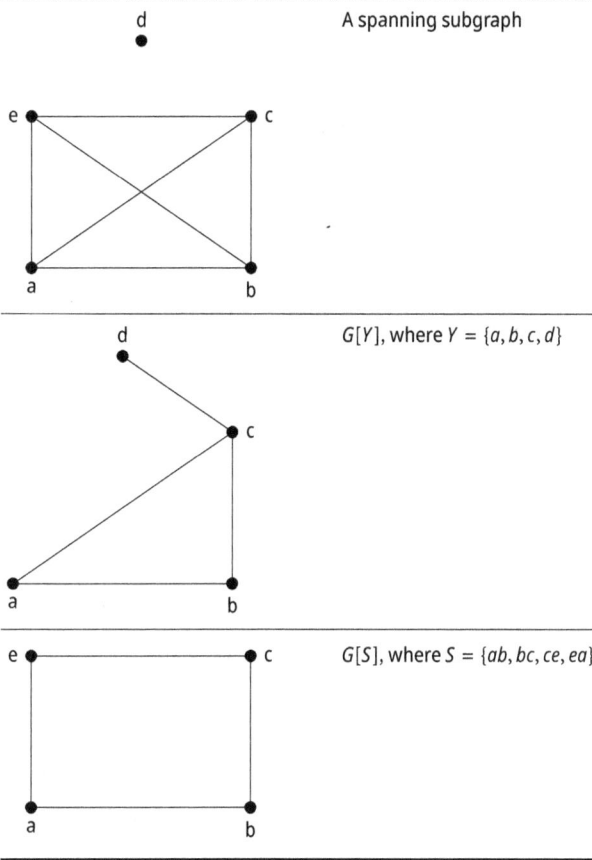

There are two processes by which new graphs are obtained from a given graph G through deletion or addition of edges:

- *Edge deletion:* $G - e$ denotes the result of deleting the edge e from G.
- *edge addition:* $G + e$ denotes the result of adding the edge e to G.

The *disjoint union* (or simply *union*) of graphs $G_1, \ldots, G_k$ with disjoint vertex sets $V_1, \ldots, V_k$ and edge sets $E_1, \ldots, E_k$, denoted by $\bigcup_{i=1}^{k} G_i$, is the graph whose vertex set is $\bigcup_{i=1}^{k} V_i$ and edge set is $\bigcup_{i=1}^{k} E_i$. A *decomposition* of a graph G is a set of subgraphs $H_1, \ldots, H_k$ that partition the edges of G, that is, $\bigcup_{i=1}^{k} E(H_i) = E(G)$ and $E(H_i) \cap E(H_j) = \emptyset$ for $1 \le i < j \le k$. A *complete n-graph* is the graph on n vertices denoted by $K_n := ([n], E^n)$, where E^n is the set of all $\binom{n}{2} := \frac{n(n-1)}{2}$ edges (i, j) for $i = 1, \ldots, n, j = i+1, \ldots, n$. Any graph $G = ([n], E)$ on n vertices is a subgraph of K_n. The complement of G is $G^c := ([n], E^n \backslash E)$. (For any two subsets P, Q of a given set R, $P\backslash Q$ stands for all elements in P that are not in Q. Note that $P\backslash Q$ may be the empty set $\emptyset$.) The *empty n-graph* $K_n^c := ([n], \emptyset)$ is the complement of K_n. $K_1 = K_1^c$ is called the *trivial* graph, that is, K_1 consists solely of a vertex with no edges. A *clique* of a graph G is a complete subgraph of G. A *k-clique* is a clique of order k, that is, a clique with k vertices. Since $k = 0, 1, 2$ covers only trivial scenarios, we always assume that $k \geq 3$ (0-cliques correspond to the empty set (sets of zero vertices), 1-cliques correspond to vertices, and 2-cliques to edges.)

Two vertices $a, b \in V$ are called *adjacent* if $(a, b) \in E$. Let $G = (V, E)$ be a graph, and let $v \in V$. Then $\Gamma(v)$ denotes the *neighborhood* of v, i.e., the set of all adjacent vertices of v. An edge is called *incident* to a vertex if the vertex is one of the edge endpoints. The *degree* of a vertex v is defined as $\deg(v) := |\Gamma(v)|$, i.e., the number of neighbors of v. A vertex v is called *isolated* if $\deg(v) = 0$. Let $|V| = n$. Then the *degree sequence* of G is the sequence of the degrees of all the vertices of G arranged in decreasing order: $\deg(v_1) \geq \deg(v_2) \geq \cdots \geq \deg(v_n) \geq 0$. Since every edge in G is connected to two distinct vertices, it follows that

$$\sum_{v \in V} \deg(v) = 2|E|. \tag{2.1.1}$$

In other words, the sum of the degrees of all vertices in a graph is equal to twice the number of edges in that graph. This is called *handshaking lemma* or *degree-sum formula* in graph theory. From the handshaking lemma it immediately follows that the number of vertices with odd degrees in a graph must be an even number. (Why?)

Paths, walks, trails, and connections. A *path P* of *length* $l \geq 1$ in a graph G is a set of vertices $V(P) = \{v_0, v_1, \ldots, v_l\}$ such that $(v_{i-1}, v_i) \in E$ for $i = 1, \ldots, l$ and $v_i \neq v_j$ for $i \neq j$; P is called an *even (odd)* path if l is even (odd). Assuming that P starts at v_0 and ends at v_l, it is called a $v_0 - v_l$ path. A *walk P* of *length* $l \geq 1$ in G is a set of vertices $V(W) = \{v_0, v_1, \ldots, v_l\}$ such that $(v_{i-1}, v_i) \in E$ for $i = 1, \ldots, l$. W is a *closed walk* if $v_l = v_0$. A *spanning walk* of G is a walk that contains all the vertices of G. A *closed spanning walk* is a closed walk that is spanning as well. A walk W is called a *trail* if all its edges $(v_0, v_1), (v_1, v_2), \ldots, (v_{l-1}, v_l)$ are distinct. A closed trail is called a *circuit*. A circuit W is called a *cycle* if $v_i \neq v_j$ for $0 \leq i < j \leq l - 1$. Note that the closed walk $\{i, j, i\}$ for $1 \leq i < j \leq n$ in K_n is not a cycle. We call such a closed walk a *semicycle*. Thus any cycle in an undirected graph has length at least 3.

Historical note. Leonhard Euler (Figure 2.1) proved the handshaking lemma while working on the seven bridges of Königsberg, asking for a walking tour of the city of Königsberg (now Kaliningrad) crossing each of its seven bridges once. Euler translated this problem into graph theory, where an Euler circuit (see Definition 2.17) of a connected graph represents the city and its bridges. Euler's insights focused on the number of odd vertices, constrained by the handshaking lemma to be even. If this count is zero, then an Euler tour exists; if it is two, then an Euler path also exists. For the seven bridges of Königsberg, with four odd vertices, neither an Euler path nor tour was possible. In the Christofides–Serdyukov algorithm for the traveling salesperson problem, the degree-sum formula's geometric implications play a vital role, allowing the algorithm to pair vertices and construct a graph where an Euler tour serves as an approximate solution for the traveling salesperson problem.

A prevalent theme in graph theory involves breaking down a structure into its constituent substructures. This can entail decomposing a closed walk into substructures such as cycles and semicycles (Proposition 2.2) or an undirected graph into connected components (Proposition 2.3). As we mentioned, the subsequent two propositions follow this theme, and their proofs are left for the reader to explore.

Proposition 2.2. *Let W be a closed walk on an undirected graph G. Then the edges of W can be decomposed into a union of the edges of cycles and semicycles.*

Let $u, v \in V$ be two distinct vertices. Then u is *connected* to v if there exists a path P starting at u and ending at v. We write $u \sim v$ if u is connected to v. Clearly, $u \sim v$ if and only if $v \sim u$. Thus $u \sim v$ if u and v are connected. It is convenient to assume that $u \sim u$, i. e., u is connected to itself. Then $\sim$ is an equivalence relation on V. (See Appendix A.1 for the definition and theorems concerning equivalence relations and classes.)

An undirected graph G is called *connected* if for all $u, v \in V$, u is connected to v or, in other words, for all $u, v \in V$, there exists a $u - v$ path. G is called *disconnected* is it is not connected.

Notice that *"connected"* serves as an adjective exclusively used in reference to graphs and pairs of vertices. However, it is not appropriate to describe a single vertex u as "disconnected." Although the expression "u is connected to v" is convenient in writing proofs, it is essential to elucidate the distinction between adjacency and connection.

Proposition 2.3. *Let $G = (V, E)$ be an undirected graph. Then there is a unique decomposition of V to a disjoint union of nonempty subsets of vertices $V = \cup_{i=1}^{k} V_k$ (where the uniqueness is up to a permutation of $V_1, \ldots, V_k$) such that the following conditions hold:*
(a) $G = \cup_{i=1}^{k} G(V_i)$.
(b) *Any two vertices in each $G(V_i)$ are connected.*

In the above setting, $G(V_1), \ldots, G(V_k)$ are called the *connected components* of G.

! **Remark 2.4.** It is stated that a graph possesses a *giant component*, indicating a connected subgraph that includes a substantial portion of the vertices. The determination of a "substantial portion" is subjective and relies on the specific problem at hand. In certain contexts, for example, $\frac{1}{10}$ may denote a large proportion, whereas in other situations a fraction of $\frac{7}{10}$ might be deemed small. The concept of a giant component often surfaces in the realm of random graph theory, a topic that will be further explored in Chapter 3.

Forests and trees. An undirected graph $G = (V, E)$ is called *acyclic* if it does not have a cycle. An acyclic and connected graph G is called a *tree*. An acyclic graph G is called a *forest*. A subgraph $T = (V, W)$ of a connected graph G with $W \subset E$ is called a *spanning tree* if T is a tree, that is, a spanning tree is a spanning subgraph that is a tree as well.

Example 2.5. Let G be the graph as in Example 2.1. Then:
- G is connected.
- G is not acyclic as it contains cycles; for example, $\{ec, cd, de\}$ is a cycle in G.
- G is not a tree because it is not acyclic.
- The following subgraph of G is a forest:

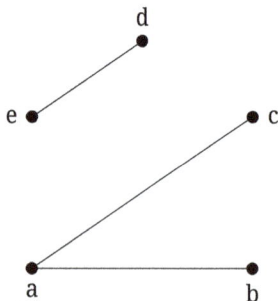

- The following subgraph of G is a spanning tree:

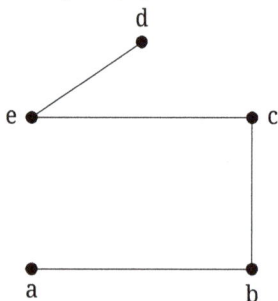

Let $G = (V, E)$ be a connected undirected graph. For $u, v \in V$, we define the *distance* $\mathrm{dist}(u, v)$ as the shortest length of a path connecting u to v. In case $u = v$, we simply make the convention $\mathrm{dist}(u, v) = 0$.

Proposition 2.6. *Let G be a connected undirected graph. The function*

$$\text{dist} : V \times V \to [0, \infty)$$

is a distance function on V. That is, (V, dist) is a metric space.

Proof. Clearly, by the definition of dist, we have $\text{dist}(u, u) = 0$ and $\text{dist}(u, v) > 0$ for all $u \neq v$ (positivity). Moreover, clearly, $\text{dist}(u, v) = \text{dist}(v, u)$ (symmetry). Note that if W is a walk of length l from u to v, then $l \geq \text{dist}(u, v)$. Equality holds if and only if W is a shortest path from u to v. Hence the distance function satisfies the triangle inequality: $\text{dist}(u, w) \leq \text{dist}(u, v) + \text{dist}(v, w)$ for all $u, v, w \in V$. Thus $\text{dist}(\cdot, \cdot)$ is a distance function on V. With the definition of the distance between vertices at hand, we can now proceed to define the diameter of a graph. The *diameter* of a graph G is defined as

$$D(G) = \max_{u,v \in V} d(u, v).$$

Assuming that G is not a tree, there exists a related notion to $D(G)$, called the *girth* of G, denoted by g_G (or simply g when there is no ambiguity regarding G). The girth is defined as the length of the shortest cycle in G. The following proposition studies the correlation of girth and diameter and is named after Eliakim Hastings Moore (1862–1932) an American mathematician known for his contributions to algebra, graph theory, and geometry. □

Proposition 2.7 (Moore's inequality). *For a graph G that is not a tree, $g \leq 2D(G) + 1$.*

Proof. Assume on the contrary that $g > 2D(G) + 1$. Clearly, G has a cycle C of length g. Therefore we may find vertices u and v on C that are connected by a path of length $D(G) + 1$. Since $D(G)$ is the diameter, there must be a path between u and v of length at most $D(G)$. However, these two paths, being distinct, will form a cycle of length at most $2D(G) + 1$, which is less than g. This contradicts the fact that g is a girth. □

In the following three propositions, we establish that every connected graph possesses at least one spanning tree. We will investigate such a spanning tree. Moreover, we will show that a connected graph may have more than one spanning tree. Nevertheless, any spanning tree of a connected graph will always have the same number of edges.

Proposition 2.8. *Let $G = (V, E)$ be a connected undirected graph. Then G has a spanning tree.*

Proof. If G is trivial, then G is its spanning tree. Assume that $n = |V| > 1$. Let $V_0 := \{v_0\}$. For each $i \in \mathbb{N}$, let $V_i := \{v \in V : \text{dist}(v_0, v) = i\}$. Then there exists a positive integer k such that $V_i \neq \emptyset$ for $i = 1, \ldots, k$ and $V_i = \emptyset$ for $i > k$. By the definition $V_i \cap V_j = \emptyset$ for $i \neq j$. Since G is connected, $V = \cup_{i=0}^{k} V_i$. Also, $E \cap (V_i \times V_j) = \emptyset$ for $j - i \geq 2$. Now let $E_i = E \cap (V_{i-1} \times V_i)$ for $i = 1, \ldots, k$. By the definitions of $V_i, i = 0, \ldots, k$, $E_i \neq \emptyset$ for $i = 1, \ldots, k$. Then $T := (V, \cup_{i=0}^{k} E_i)$ is a spanning tree of G. □

The tree described in the proof is called a *rooted* tree with root v_0. It is straightforward to show the following:

Proposition 2.9. *The following statements are equivalent for an undirected graph* $G = (V, E)$:

(a) *G is a tree.*

(b) *G is a minimal connected graph on V, that is, G is connected, and for any edge $(u, v) \in E$, the subgraph $H = (V, E\backslash\{(u, v)\})$ is disconnected. (This is equivalent to saying that every edge of G is a bridge.) (See Problem 2.4-3.)*

(c) *G is a maximal acyclic graph, that is, G is acyclic, and for any $u \neq v \in V$ such that $(u, v) \notin E$, the graph $H = (V, E \cup \{(u, v)\})$ contains a cycle.*

Proposition 2.10. *A tree of order n has size $n - 1$. A forest of order n with k components has size $n - k$.*

Proof. Assume first that $G = (V, E)$ is a tree. We show that $|E| = |V| - 1$ by induction on $n := |V|$. For $n = 1$, $G = K_1$ and $E = \emptyset$. Assume that the claim holds for all trees of order at most m. Let $G = (V, E)$ be a tree with $|V| = m + 1$. Let $(u, v) \in E$ and consider the subgraph $H = (V, E\backslash(u, v))$. Then G is a union of two disjoint trees $G_1 = (V_1, E_1)$ and $G_2 = (V_2, E_2)$, each of order at most m. The induction hypothesis yields that $|E_1| = |V_1| - 1$, $|E_2| = |V_2| - 1$. Hence $|E| = |E_1| + |E_2| + 1 = |V_1| + |V_2| - 1 = |V - 1|$.

To prove the corresponding claim for the forest, use the fact that a forest with k components is a disjoint union of k trees. ☐

Corollary 2.11. *Any tree of size 2 at least has at least two vertices of degree 1.*

Proof. Let G is a tree of order $n \geq 2$. Let $d_1 \geq \cdots \geq d_n$ be the degree sequence of G. Since G is connected, $d_n \geq 1$. Since the size of G is $n - 1$, it follows that $\sum_{i=1}^{n} d_i = 2(n - 1)$. Since each $d_i \in \mathbb{N}$, it follows that $d_{n-1} = d_n = 1$. Otherwise, if $d_{n-1} \geq 2$ and $d_n \geq 1$, then we will have that $\sum_{i=1}^{n} d_i \geq 2(n - 1) + 1 > 2(n - 1)$, which contradicts the above equality. ☐

Notice that a vertex in a tree of size at least 2 is called a *leaf* if its degree is 1. A tree of size at least 2 that has exactly two leaves is a path $\bullet - \bullet - \cdots - \bullet - \bullet$. A tree of size $n \geq 2$ that has $n - 1$ leaves is called a *star*. It belongs to the class of the so-called complete bipartite graphs, which will be further explored later in this chapter.

k-Colorability. A *proper coloring* of an undirected G graph is an assignment of colors to the vertices of a graph in such a way that no two adjacent vertices share the same color. A *k-coloring* of G is a proper coloring that involves a total of k colors. A graph that has a k-coloring is said to be *k-colorable*. The *chromatic number* of G, denoted by $\mathcal{X}(G)$, is defined as the minimum number of colors required in a proper coloring of that graph.

Example 2.12. Consider the following graph G:

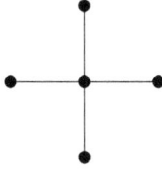

Then the following is a proper coloring of G:

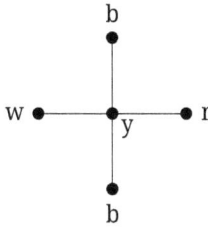

b: blue
r: red
w: white
y: yellow

Clearly, this proper coloring is a 4-coloring. Thus G is 4-colorable. On the other hand, G can be colored with fewer number of colors. The following coloring for G suggests $\mathcal{X}(G) = 2$:

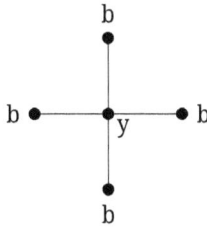

Planar graphs. An undirected graph G is called *planar* if it can be drawn on a plane without any edges crossing. Practically, imagine being able to trace the edges of the graph without ever lifting your pen from the paper. Note that the definition of a planar graph incorporates the phrase "*it can be drawn.*" This means that even if a graph does not initially appear to be planar, there exists the possibility that it is. It may be feasible to redraw the graph in a manner where no edges intersect.

Example 2.13. Consider the following graph G:

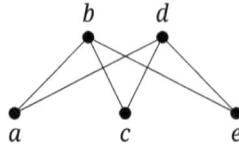

At a glance, it seems to us that G is not a planar graph. However, G can be redrawn as follows entailing that it is a planer graph:

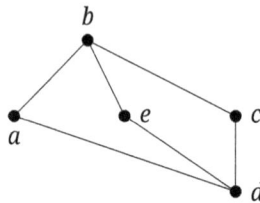

However, when experimenting with drawing of K_5, there does not seem to be any way to eliminate the crossings:

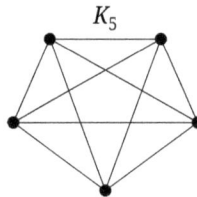

When a planar graph is depicted without any crossing edges, it partitions the plane into a set of regions known as *faces*. Each face is enclosed by a closed walk referred to as the *boundary* of the face. As a convention, we consider the unbounded area outside the entire graph as one face. The *degree* of this face corresponds to the length of its boundary. For example, consider the graph in Example 2.13 (the redrawn version). Then the graph has three faces:

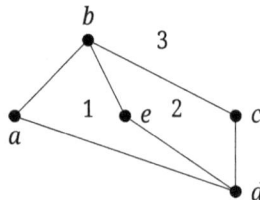

The boundary of face 1 has edges ba, bd, de, and db, so this face has degree 4. The boundary of face 2 has edges be, ed, dc, and cb, so this face has degree 4 as well. The boundary of face 3 (the unbounded face) has edges ba, ad, dc, and cb, so this face has degree 4 too!

Suppose we have a connected planar graph with f faces of order n and size m. Then we have Euler's formula $n - m + f = 2$. The proof is left as Problem 2.4-7.

Bipartite graphs. A *bipartite* graph is a graph where the vertices can be divided into two disjoint sets such that all edges connect a vertex in one set to a vertex in another set. In other words, $G = (V, E)$ is bipartite if V decomposes into two disjoint nonempty sets V_1 and V_2 such that $E \subset V_1 \times V_2$. The sets V_1 and V_2 are called the *partite sets* of G. In what follows, we first make a few obvious observations about bipartite graphs and then state and prove a theorem in which bipartite graphs are classified based on the nature of their cycles.

-   A graph is bipartite if and only if each of its connected components is bipartite.
-   Every subgraph of a bipartite graph is bipartite.
-   No odd cycle is bipartite.
-   Every tree is bipartite.

Theorem 2.14. *Let G be an undirected graph. Then G is bipartite if and only if G has no odd cycles.*

Proof. Clearly, G is bipartite if and only if each connected component is bipartite. Thus it suffices to prove the theorem in the case $G = (V, E)$ is connected. If G is bipartite, then any walk is of the form $V_1 - V_2 - V_1 \cdots$ or $V_2 - V_1 - V_2 \cdots$. So any closed walk on G has an even length. In particular, any cycle has an even length.

Conversely, assume that any cycle, if it exists, is even. Since the length of a semicycle is 2, Proposition 2.2 yields that any closed walk in G has an even length. Hence any two walks between any two vertices $u, v \in V$ have the same parity. Fix $v_0 \in V$. Let V_1 be the set of all vertices in $v \in V$ such that any walk from v_0 to v has an odd length. Then $V_2 := V \backslash V_1$ is the set of all the vertices $v \in V$ such that any walk from v_0 to v has an even length. Hence $E \subset V_1 \times V_2$. □

Having Theorem 2.14 at hand together with the definition of a bipartite graph, we have the following techniques for checking whenever a graph is bipartite:
-   If it is bipartite, then prove it by finding two independent sets.
-   If it is not bipartite, then find an odd cycle.

Definition 2.15. Let U and V be two disjoint sets of vertices of cardinalities m and n, respectively. The complete bipartite graph $K_{m,n}$ has order $m + n$ and size mn and is obtained by joining every vertex $u \in U$ to every vertex $v \in V$. The graph $K_{m,n}$ is isomorphic to $G := (U \cup V, U \times V)$. Note that star $K_{1,n-1}$ is a complete bipartite graph on $|U| = 1$, $|V| = n - 1$.

Example 2.16. In the following, we see a few complete bipartite graphs:

$K_{1,3}$ $K_{2,3}$ $K_{3,3}$

$K_{4,4}$

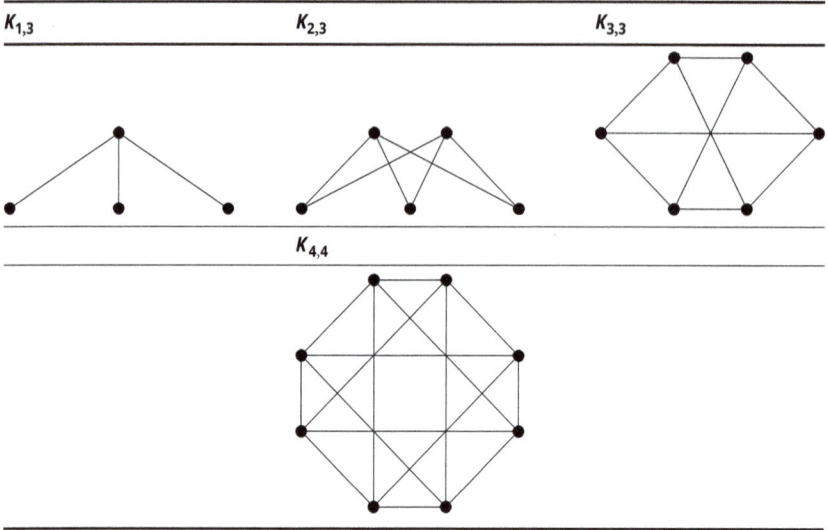

Definition 2.17. Let $G = (V, E)$ be an undirected connected graph.
– C is called a Hamiltonian cycle in G if C is a cycle on all vertices of G.
– G is called a Hamiltonian graph if G possesses a Hamiltonian cycle.
– P is called a Hamiltonian path in G if P is a path on all vertices of G.
– C is called Eulerian circuit if C is a circuit of G containing all edges of E.
– T is called an Eulerian trail if T is a trail that contains all the edges of G.
– G is called an Eulerian graph if it has an Eulerian trail.

Example 2.18. Consider the following graph G:

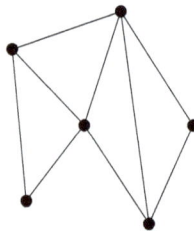

We extract the following Hamiltonian cycle from G (in blue):

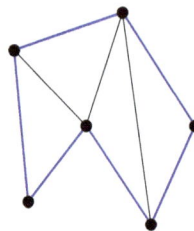

Therefore G is a Hamiltonian graph. Moreover, we extract the following Hamiltonian path from G (in blue):

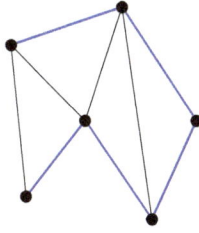

We also extract an Eulerian trial from G, 1231435465:

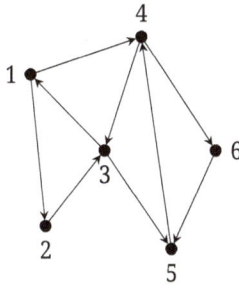

We may verify that G has no Eulerian circuit by trial and error. However, it is evident that trial and error is not an optimal way to determine the existence of Eulerian circuits. In Theorem 2.19, we will provide a practical way to investigate the existence of Hamiltonian circuits. Theorem 2.19 may also be used to determine the existence of nonclosed Eulerian trials.

Theorem 2.19. *Let $G = (V, E)$ be an undirected connected nontrivial graph. Then:*
1. *G has an Eulerian circuit if and only if the degree of each vertex in G is even.*
2. *G contains a nonclosed Eulerian trail if and only if G has two vertices of odd degrees and all other vertices have even degrees.*

Proof. When traveling on an Eulerian circuit, every time you enter the vertex v, you exit it, from which it follows that the degree of each vertex is even. We prove by induction that on the size $m = |V|$, if all the vertices of G have even degrees, then G contains an Eulerian circuit. For $m = 3$, $G = K_3$, and the unique cycle on K_3 is Eulerian. Assume that the theorem holds for any G of size $m \geq 3$. Assume that $|E| = m + 1$. Then $|V| > 3$. Since each vertex of v has an even degree, G cannot be a tree, and hence there exists a cycle C in G. Consider a subgraph of $G_1 = (V, E_1)$ obtained from G by deleting all edges in the cycle C. Suppose first that G_1 is connected. Then the induction hypothesis implies that G_1 has an Eulerian circuit C_e, which can be started from any vertex $v \in V$. Start it from a vertex v in the cycle C. Complete first C_e ending at v and then complete the cycle C to obtain an Eulerian circuit on G. If G_1 is disconnected, then any vertex in any nontrivial connected component H has even degree. The induction hypothesis yields that each nontrivial component H has an Eulerian circuit. It is straightforward to show how to combine these Eulerian circuits with C to obtain a Eulerian circuit on G.

Assume that G contains a nonclosed Eulerian trail T. Then the initial and the end vertex T have odd degrees, and all other vertices have even degrees. Suppose that a connected G has two vertices u, v of odd degrees and all other vertices of V have even degrees. Let P be a path from u to v in G. Let $G_1 = (V, E_1)$ be the subgraph of G obtained by deleting all edges on the path P. Then each vertex $v \in V$ has an even degree in G_1. Hence every nontrivial component of G_1 has an Eulerian circuit. Combine these Eulerian circuits with the path P to obtain a nonclosed Eulerian trail from u to v. □

Computational complexity of a given property on graphs. Let $\mathbf{P}$ be a graph property. For example: $\mathbf{P}_E$ – a graph G has an Eulerian circuit; $\mathbf{P}_H$ – a graph G has a Hamiltonian cycle. How "difficult" is to find out if G has the property $\mathbf{P}$? By difficult we mean the computational complexity of $\mathbf{P}$, i. e., how many computational operations we need in the *worst case scenario* to find if a given G has the property $\mathbf{P}$. Clearly, the amount of computation depends on the size n of G.

Definition 2.20. A property $\mathbf{P}$ of a graph G is called polynomial if there exists a polynomial $p(x) = a_0 x^l + a_1 x^{l-1} + \cdots + a_l$ such that for every $G = ([n], E)$, we need at most $p(n)$ operations to find out if G has the property $\mathbf{P}$ for all $n \in \mathbb{N}$. The set of all polynomial properties $\mathbf{P}$ of graphs is denoted by **Po**.

A property $\mathbf{P}$ is called nondeterministic polynomial if there exists a polynomial $q(x) = b_0 x^m + b_1 x^{m-1} + \cdots + b_m$ with the following property. For every $G = ([n], E)$ that is claimed by somebody to have the property $\mathbf{P}$ with the provided documentation of the claim, it will take you at most $q(n)$ steps to find if the person is right. The set of all nondeterministic polynomials $\mathbf{P}$ is denoted by **NP**.

The property $\mathbf{P}_E$ is a polynomial property in view of Theorem 2.19. The property $\mathbf{P}_H$ is nondeterministic polynomial. Indeed, to substantiate a claim that a given graph $G = ([n], E)$ has a Hamiltonian cycle, we should supply the Hamiltonian cycle C. Now it is your task to check out if C is indeed a Hamiltonian cycle in G as the person claims. This check is clearly polynomial in n.

It is straightforward to show that $\mathbf{Po} \subseteq \mathbf{NP}$. It is not known if $\mathbf{Po} \subsetneq \mathbf{NP}$. (There is a one million dollar prize for the solution of this problem!) It is believed that $\mathbf{NP}$ is strictly larger than $\mathbf{Po}$. It is known that $\mathbf{NP}$ contains a subset of the most difficult problems called *NP-complete*, which are denoted by $\mathbf{NPC}$. Every two problems in the class $\mathbf{NPC}$ are polynomially equivalent. Put it differently, if $\mathbf{P} \in \mathbf{NPC}$ is polynomial, then $\mathbf{Po} = \mathbf{NPC}$. Roughly speaking, any problem in $\mathbf{NPC}$ is believed to have an exponential complexity. It is known that $\mathbf{P}_H \in \mathbf{NPC}$.

2.2 Directed graphs (digraphs)

Warm up. A *directed* graph, sometimes referred as *digraph*, is $G := (V, E)$, where V is the set of vertices, and $E \subset V \times V$ is the set of *directed* edges. Formally, $G = (V, E)$ is a digraph where

1. V is a set of vertices (or nodes), and
2. E is a set of ordered pairs of vertices, called *arcs* (or *directed* edges or *arrows*).

When there is no ambiguity, we simply refer to a directed edge as an edge. An edge of the form $(u, v) \in E$ with $u \neq v$ is called the *edge from u to v*. In this case, we say that (u, v) is an *outgoing* edge of the vertex u. We also say that (u, v) is an *incoming* edge of v. An edge of the form $(u, u) \in E$ for some $u \in G$ is called a *loop* (on u). A digraph without loops is called *loopless*. Throughout, we assume that $G = (V, E)$ is a digraph, unless stated otherwise. A digraph G is called an *oriented* graph if G has no opposite pairs of directed edges, meaning that at most one of (u, v) and (v, u) may be an edge of G. The digraph $G_1 = (V_1, E_1)$ is called a *subgraph* of a digraph G if $V_1 \subset V$ and $E_1 \subset E$. We highlight the following families of subgraphs:

-   *vertex-induced subgraph:* Let $V_1 \subset V$. Then $G(V_1) := (V_1, E \cap (V_1 \times V_1))$ is called the *vertex-induced subgraph* (or simply *induced subgraph*) of G by V_1.
-   *edge-induced subgraph:* Let $E_1 \subset E$. Let $V_1 \subset V$ be all the vertices V that appear in the edges E_1. Then $G(E_1) := (V_1, E_1)$ is the *edge-induced subgraph* (or simply *induced subgraph*) of G by E_1.

A *directed path* (or more simply, a *path*) P in G is given by the vertices $V(P) = \{v_0, v_1, \ldots, v_l\}$ where $v_i \neq v_j$, for $i \neq j$ and $(v_{i-1}, v_i) \in E$ for $i = 1, \ldots, l$. A *directed trail* (or a *trail*) T in G is given by the vertices $V(T) = \{v_0, v_1, \ldots, v_l\}$, where $(v_{i-1}, v_i) \in E, i = 1, \ldots, l$, are l distinct edges in G. A *directed cycle* (or a *cycle*) C of length l is given by a trail T as above such that $v_l = v_0$ and $v_i \neq v_j$ for $1 \leq i < j \leq l$. A cycle of length 1 is a loop on v_0, and a cycle of length 2 is given by two edges (v_0, v_1) and (v_1, v_0). A *directed walk* (or a *walk*) W of length $l \geq 1$ is given by $V(W) = \{v_0, v_1, \ldots, v_l\}$, where $(v_{i-1}, v_i) \in E, i = 1, \ldots, l$.

For $v \in V$, let

$$\deg_{\text{out}}(v) := \left|\{u \in V : (v, u) \in E\}\right| \quad \text{and} \quad \deg_{\text{in}}(v) := \left|\{u \in V : (u, v) \in E\}\right|$$

be the *out* and *in* degrees of the vertex v. Note that if the loop $(v, v) \in E$, then it contributes to out and in degrees of v.

Example 2.21. Consider the following digraph G:

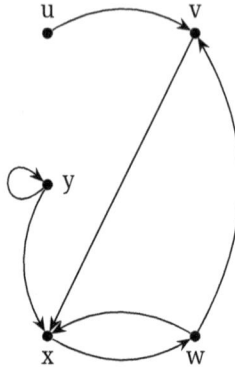

Then the vertex set of G is $V = \{u, v, w, x, y\}$, and its directed edge set is

$$E = \{(y,y), (y,x), (x,w), (w,v), (u,v), (w,x), (v,x)\}.$$

Furthermore, $y \to y$ is a loop, $x \to w \to x$ is a cycle of length 2, and $x \to w \to v \to x$ is a cycle of length 3. Also,

$$\deg_{\text{out}}(u) = 1, \quad \deg_{\text{out}}(v) = 1, \quad \deg_{\text{out}}(w) = 2, \quad \deg_{\text{out}}(x) = 1, \quad \deg_{\text{out}}(y) = 1,$$
$$\deg_{\text{in}}(u) = 0, \quad \deg_{\text{in}}(v) = 2, \quad \deg_{\text{in}}(w) = 1, \quad \deg_{\text{in}}(x) = 2, \quad \deg_{\text{in}}(y) = 1.$$

It is evident that the sum of in degrees equals the sum of out degrees.

Proposition 2.22. *Let $G = (V,E)$ be a digraph. Then*

$$\sum_{v \in V} \deg_{\text{out}}(v) = \sum_{v \in V} = \deg_{\text{in}}(v) = |E|.$$

Proof. It is straightforward and is left as an exercise (Problem 2.4-14). □

Directed graph connectivity. We begin with obtaining digraphs from a given undirected graph. If $H = (V, E_{\text{undir}})$ is an undirected graph, then there are two standard ways to associate with H a digraph $G = (V, E)$ on the same set of vertices V:

- The *maximal directed graph* $G = (V, E_{\text{max}})$ of H is given by letting (u, v) and (v, u) be in E_{max} if and only if the indirected edge uv is in E_{undir}.
- A *minimal directed graph* $G = (V, E_{\text{orient}})$ where $E_{\text{orient}} \subset E_{\text{max}}$ is such that if $uv \in E_{\text{undir}}$, then either $(u, v) \in E_{\text{orient}}$ or $(v, u) \in E_{\text{orient}}$ but not both. This is equivalent to an orientation of any undirected edge $uv \in E_{\text{undir}}$, that is, every minimal directed graph is an oriented graph.

Conversely, any digraph $G = (V, E)$ induces the following unique undirected graph $H = (V, E_{\text{undir}})$ on the same set of vertices: For $u, v \in V$, $uv \in E_{\text{undir}}$ if and only $u \neq v$ and either (u, v) or (v, u) are in E (or both).

A digraph $G = (V, E)$ is called *weakly connected* (or, more simply, *connected*), if the induced directed graph $H = (V, E_{\text{undir}})$ is connected. Otherwise, G is called *disconnected*, that is, G is disconnected if there exists a decomposition of V into disjoint nonempty subsets $V = V_1 \cup V_2$ such that $G = G(V_1) \cup G(V_2)$. For each digraph $G = (V, E)$, there exists a decomposition $V = \cup_{i=1}^{k} V_k$ into nonempty disjoint sets, called *connected components* of G, such that $G(V) = \cup_{i=1}^{k} G(V_i)$, where each $G(V_i)$ (a component of G), is connected. This decomposition is unique up to permutation of $V_1, \ldots, V_k$. We make the convention that a digraph with one vertex only is assumed to be connected.

A digraph G is called *strongly connected* if for any two distinct vertices $u, v \in G$, there exists a path P from u to v, i.e., $V(P) = \{u_0 = u, u_1, \ldots, u_l = v\}$. A *maximal strongly connected subgraph H* of G is a strongly connected subgraph of it, that is, it is maximal with respect to the property of strong connectedness. (H is not properly contained in any subgraph of G.)

It is evident to note the following about properties about strong connectivity:

- Every strongly connected digraph is connected.
- Not all connected graphs are strongly connected. For example, a digraph having two vertices and exactly one edge from one vertex to another is connected but not strongly connected.
- If G is the maximal directed graph of an undirected graph H, then G is strongly connected if and only if H is connected.
- Every closed path on a digraph is strongly connected.

Example 2.23. Consider the following digraph $G = (V, E)$:

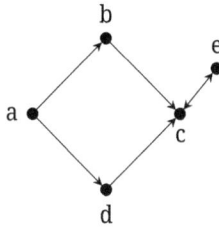

Then G induces the following undirected graph $H = (V, E_{\text{undir}})$:

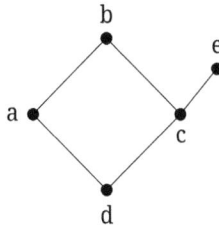

Clearly, G is connected because H is connected. However, G is not strongly connected because there is no directed path from b to a.

Example 2.24. Consider the following undirected graph $H = (V, E_{undir})$:

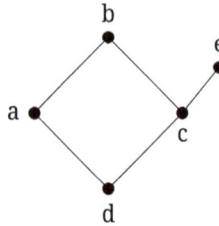

a) The associated maximal directed graph $G = (V, E_{max})$ is

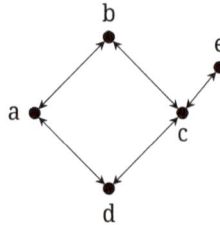

Since H is connected, then G is strongly connected.

b) The associated minimal digraph $G = (V, E_{orient})$ is

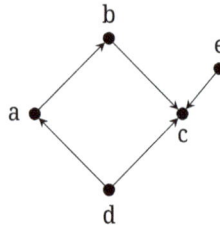

Clearly, G is connected but not strongly connected. For example, there is no directed path from e to d.

Remark 2.25. Given an undirected graph, we explored the derivation of two types of digraphs from it, maximal and minimal digraphs. It is worth noting that we can also associate an undirected graph with a given digraph. This undirected graph is referred to as the *underlying graph*, obtained by removing all edge directions.

If an undirected graph has no cycles, then it is classified as a tree or a forest. However, the appearance of a directed graph without cycles differs. Consider, for example, the graph G shown in Example 2.23. Although it is connected and lacks directed cycles, it does not resemble a tree.

Definition 2.26. A directed graph G is called *acyclic* if G has no directed cycles.

We may argue that G is acyclic if and only if it has no closed-directed walks. (Why? See Problem 2.4-15.) Acyclic digraphs might not seem particularly intriguing. However, initial impressions can be deceiving. In reality, they play a significant role in numerous optimization problems, boasting several noteworthy properties. In Chapter 4, we will utilize acyclic digraphs frequently.

Proposition 2.27. *Let $G = (V, E)$ be an acyclic digraph. Then:*
1. *V decomposes into a disjoint union of k nonempty subsets $V_1, \ldots, V_k$ with the property that for each $v \in V_1$, $\deg_{in}(v) = 0$, i. e., v has no incoming edges.*
2. *If $k \geq 2$, i. e., G contains nonisolated vertices, then for each $2 \leq i \leq k$, V_i consists of all u in V such that:*
   (a) *There exist $v(u) \in V_1$ and a path of length $i - 1$ connecting $v(u)$ to u.*
   (b). *If there exists a path from $v \in V_1$ to u, then its length is at most $i - 1$.*

Proof. We first show that V_1 is nonempty. Take any vertex $v_0 \in V$. If $\deg_{in}(v_0) = 0$, then $v_0 \in V_1$. If $\deg_{in}(v_0) > 0$, then there exists $v_{-1} \in V$ such that $(v_{-1}, v_0) \in E$. Suppose we can continue this process j steps, i. e., we have a walk W on G such that $V(W) = \{v_{-j}, \ldots, v_{-1}, v_0\}$. Since G is acyclic, W is a path. Suppose that $|V| = n$. Then $j \leq n - 1$. So this process must stop at some $j \leq n - 1$, i. e., $v_{-j} \in V_1$.

Denote by V_1 the set of all $v \in V$ such that $\deg_{in}(v) = 0$. Assume that $V_1 \neq V$. Let $v_0 \in V \backslash V_1$, i. e., v_0 is any vertex in V that is not in V_1. By the above arguments there exists a path W on G such that $V(W) = \{v_{-j}, \ldots, v_{-1}, v_0\}$ and $v_j \in V_1$. Let i be the maximal value of all possible values of j. Then v_0 belongs to V_{i+1}. Since V is finite, the maximal possible value of i is k. $\qquad\square$

Definition 2.28. Let $G = (V, E)$ be acyclic digraph. A vertex $v \in V$ is called an initial vertex (or source vertex) if $\deg_{in}(v) = 0$ and is called a terminal vertex (or sink vertex) if $\deg_{out}(v) = 0$. Any nonterminal vertex is called transient.

Remark 2.29. It is worth mentioning that:
- The acyclicity of graph G guarantees the existence of at least one initial vertex and one terminal vertex.
- A vertex $v \in V$ is initial and terminal if and only if v is isolated: $\deg_{in}(v) = \deg_{out}(v) = 0$.
- Every nonclosed directed path has exactly one initial vertex and one terminal vertex.

Example 2.30. Let G be the following digraph on three vertices [3] and the edges $E = \{(1, 2), (1, 3), (2, 3)\}$. Then G is acyclic; $V_1 = \{1\}$, $V_2 = \{2\}$, $V_3 = \{3\}$. The induced directed graph with the same vertices and undirected edges consists of one 3-cycle $(1, 2), (2, 3), (3, 1)$. The vertex 1 is initial, and the vertex 3 is terminal.

Assume that the conditions of Proposition 2.27 hold. Then:
- V_1 is the set of initial vertices.

- V_k is a subset of terminal vertices.
- Any V_i may contain a terminal vertex.

Structure of connected digraphs. Let $G = (V,E)$ be a connected digraph. On the set of vertices V of G, we introduce the following relation ~:

(a) $v \sim v$ for all $v \in V$.

(b) For $u, v \in V$ such that $u \neq v$, $u \sim v$ if and only if there exists a path (walk) in G connecting u to v and v to u.

It is straightforward to verify that ~ is an equivalence relation on V:

- $u \sim u$,
- $u \sim v \Longleftrightarrow v \sim u$,
- $u \sim v$ and $v \sim w \Rightarrow u \sim w$.

Hence V decomposes into a disjoint union of m nonempty equivalence classes $V = \cup_{i=1}^{m} V_i$. So for each V_i, any two members $u, v \in V_i$ are connected in G, whereas for two disjoint V_i, V_j and any $u \in V_i$, $v \in V_j$, either u is not connected to v, or v is not connected to u (or both); that is, each $G(V_i)$ is a maximal strongly connected subgraph of G. Every $G(V_i)$ is called a *strongly connected component* of G. It is evident that G is strongly connected if and only if $m = 1$.

Example 2.31. Consider the following digraph G, where two strongly connected components are highlighted (in blue):

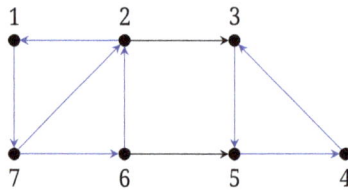

Clearly, each of the strongly connected components is a strongly connected digraph. However, G is not strongly connected. For example, there is no path connecting 3 to 2.

Given a connected digraph $G = (V,E)$ with strongly connected components $\{G(V_i)\}_{i=1}^{m}$, the associated *reduced* graph $G_{rdc} = (V_{rdc}, E_{rdc})$ is given as follows:

1. vertices $V_{rdc} = \{\{V_1\}, \ldots, \{V_m\}\}$, that is, the vertices of G_{rdc} are the equivalence classes.

2. edges follow the following two conditions:

   (a) For $i \neq j$, $(\{V_i\}, \{V_j\}) \in E_{rdc}$ if and only if there exist $u \in V_i$ and $v \in V_j$ such that u is connected to v.

   (b) G_{rdc} is loopless.

We now list the essential properties of the reduced graph:
- G_{rdc} has fewer vertices and edges than G, making it easier to analyze reachability between strongly connected components.
- Since each V_i is an equivalence class, it follows that G_{rdc} is acyclic. (Why? See Problem 2.4-17.)
- Since G is connected, it follows that G_{rdc} is connected. Thus G_{rdc} gives the exact information on the "communication" between the maximal strongly connected subgraphs of G, that is, the ability to reach one strongly connected component from another in G is accurately reflected in G_{rdc}.
- Proposition 2.27 gives the structure of G_{rdc}. The equivalence class V_i is called initial, terminal, or transient according to the status of the vertex $\{V_i\}$ in the acyclic graph G_{rdc}.
- That G_{rdc} is loopless ensures clarity in its representation.

Theorem 2.32. *Let $G = (V, E)$ be a strongly connected digraph. Let p be the gcd (the greatest common divisor) of the lengths of all cycles of G.*
1. *$p = 1$ if and only if there exists a positive integer N such that for all $u, v \in V$ and $m \geq N$, there exists a walk W on G from u to v of length m.*
2. *$p \geq 2$ if and only if the following conditions hold. It is possible to divide V into p nonempty disjoint sets $V_1, \ldots, V_p$ such that the following conditions hold:*
   - *First,*

$$E \cap (V_i \times V_{i+1}) \neq \emptyset \quad \text{for } i = 1, \ldots, p, \quad E = \cup_{i=1}^p E \cap (V_i \times V_{i+1}), \qquad (2.2.1)$$

   where $V_{p+1} = V_1$.
   - *Second, there exists a positive integer N such that for all $m \geq N$ and $u, v \in V_i$, there exists a walk from u to v in pm steps for each $i = 1, \ldots, p$.*

There is a matrix-theoretic proof of this theorem. See, for example, [32, Chapter 3]. We provide a concise proof of Theorem 2.32 using the following well-known theorem.

Theorem 2.33. *Let $0 < a_1 < \cdots < a_k$ be k positive integers whose gcd is $p \in \mathbb{N}$. Then there exists $N \in \mathbb{N}$ such so that for all $m \geq N$, there exist nonnegative integers $b_1, \ldots, b_k$ such that $mp = b_1 a_1 + \cdots + b_k a_k$.*

Proof. By considering $a_i' = \frac{a_i}{p}$, $i = 1, \ldots, k$, it suffices to prove the theorem for $p = 1$. Since for $k = 1, p = a_1 = 1$, the theorem in this case is trivial because $m = ma_1$. Let $k > 1$. If $a_1 = 1$, then again the theorem is trivial since $m = ma_1 + 0a_2 + \cdots + 0a_k$. So assume that $2 \leq a_1 < \cdots < a_k$.

Since the greatest common divisor of $a_1, \ldots, a_k$ is 1, it is known that there exists integers $c_1, \ldots, c_k$ such that

$$c_1 a_1 + c_2 a_2 + \cdots + c_k a_k = 1. \qquad (2.2.2)$$

For a_1, a_2, we apply the Euclid algorithm to find $\gcd(a_1, a_2)$. For example, for $a_2 = 3$ and $a_2 = 17$, we have $6 \cdot 3 + (-1) \cdot 17 = 1$. Apply the Euclid algorithm: $17 = 5 \cdot 3 + 2; 3 = 1 \cdot 2 + 1 \Rightarrow 17 = 5 \cdot 3 + (3-1) = 6 \cdot 3 - 1$. For $k > 2$, we first apply the Euclid algorithm to find the $g_2 = \gcd(a_1, a_2)$. Then apply the Euclid algorithm to find $g_3 = \gcd(g_2, a_3)$, and so forth.

Since $2 \le a_1 < \cdots < a_k$ and $c_1, \ldots, c_k$ are integers, we must have at least one negative integer c_i and one positive integer c_j. Let $L := \max(-c_1, \ldots, -c_k)$ and $A := \sum_{i=1}^{k} a_i$. We claim that any $m \ge LA^2$ is expressible as a nonnegative linear combination of $a_1, \ldots, a_k$ with nonnegative integer coefficients.

First, note that if m is divisible by A, i. e., $m = tA$, then $m = ta_1 + ta_2 + \cdots + ta_k$ and express m as a nonnegative linear combination of $a_1, \ldots, a_k$ with nonnegative integer coefficients. Thus it is left to consider the case where $m = tA + j$ for some $j = 1, \ldots, A - 1$. Using the assumption $m \ge LA^2$, we can assume that $t \ge LA$. Multiply (2.2.2) by j to deduce that

$$m = tA + j = t\sum_{i=1}^{k} a_i + j\sum_{i=1}^{k} c_i a_i = \sum_{i=1}^{k}(t + jc_j)a_j.$$

Since $c_j \ge -L, j \le A - 1$, it follows that $(t + jc_j) \ge LA - (A-1)L = L > 0$. ☐

Proof of Theorem 2.32. Fix a vertex $v \in V$. Consider the following closed walks in G starting and ending at v: $\mathcal{W} := \{W_1, \ldots, W_M\}$. First, consider all cycles starting and ending at v. Now consider all closed walks that van be decomposed into two cycles in G. Continue in this manner until any cycle in G appears at least in one of these walks. It is not difficult to show that the gcd of all lengths of the walks in $\mathcal{W}$ is equal to the gcd of all cycles in G.

Assume first the case $p = 1$. Theorem 2.33 yields that there exists $N_1 \in \mathbb{N}$ such that for all $m \ge N_1$, there exists a closed walk W of length m starting and ending at v that is obtained by using the set of walks in $\mathcal{W}$.

Since G is strongly connected, there exists a path of length $P(v, w)$ from u to w ($u \ne w$), of length at most $n - 1 := |V| - 1$. To get from u to w, we first go from u to v in a path $P(u, v)$ if $u \ne v$, then we take a closed walk of any length $\ge N_1$ around v, and then we take a path $P(v, w)$. This shows that we can go from any vertex u to any vertex u' in a walk of length $m \ge N_1 + 2(n - 1)$.

Assume now that the gcd of all cycles G is $p \ge 2$. Since any closed walk W in G can be decomposed into a sum of cycles, it follows that any closed walk on G is divisible by p. Let $u \ne w$. Consider two walks W_1 and W_2 from u to w. Complete each walk to a closed walk from u to itself by taking a fixed path from w to u. Since each closed walk is divisible by p, it follows that the difference of lengths of W_1 and W_2 is divisible by p.

Fix a vertex v as above. The arguments of the proof of the theorem for $p = 1$ yield that there exists N_1 such that for ally $m \ge N_1$, there is a closed walk from v to v in mp steps. For $i = 1, \ldots, p$, let V_i be all vertices in $u \in V$ such that there exists a walk of from v to u of length l, where $l - i$ is divisible by p. Note that $v \in V_p$. Next, observe that all vertices $u \in V$ with $(w, u) \in E$ for some $w \in V_p$ form exactly the set of vertices V_1. Since G is strongly connected, $\deg_{\text{out}}(v) > 0$, then $V_1 \ne \emptyset$. Now V_2 is the set of vertices $u \in V$

with $(w, u) \in E$ for some $w \in V_2$. Since G is strongly connected and V_1 is nonempty, it follows that V_2 is not empty. Continue in the same manner to deduce that $V_1, \ldots, V_{p-1}$ are not empty. Clearly, V_i is only connected to V_{i+1} for $i = 1, \ldots, p$. The rest of the theorem follows easily. $\qquad\square$

Remark 2.34. A digraph is said to be *aperiodic* if there is no integer $p > 1$ that divides the length of every cycle in the graph. Equivalently, a graph is aperiodic if the greatest common divisor of the lengths of its cycles is one (Scenario (1) in Theorem 2.32). Otherwise, G is called *periodic* (Scenario (2) in Theorem 2.32). Note that this greatest common divisor for a digraph G is called the *period* of G.

Example 2.35. Consider the following digraphs G and H:

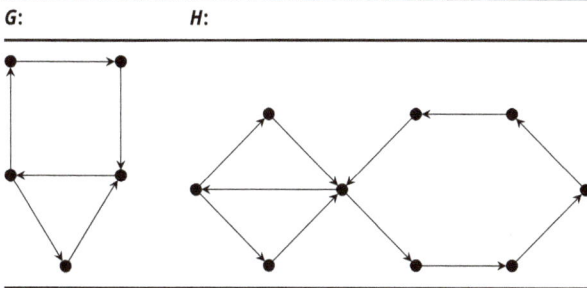

G has two cycles of lengths 3 and 4. Since $\gcd(3, 4) = 1$, G is aperiodic. On the other hand, H has three cycles of lengths 3, 3, and 6. Since $\gcd(3, 3, 6) = 3$, H is periodic with period 3.

We now briefly discuss some aspects of Theorem 2.33. It suffices to consider the case $2 \le a_1 < \cdots < a_k$ such that $a_1, \ldots, a_k$ are *coprime*, i. e., the gcd of $a_1, \ldots, a_k$ is 1. The problem of expressing $m \in \mathbb{N}$ as a linear combination of $a_1, \ldots, a_k$ with nonnegative integers is called the *coin problem*. Can we express the quantity of money m using only k types of coins of denomination $a_1, \ldots, a_k$? The largest number $f(a_1, \ldots, a_k)$ that *cannot* be expressed as a linear combination of $a_1, \ldots, a_k$ with nonnegative integers is called the *Frobenius* number of $a_1, \ldots, a_k$. For $k = 2$, Sylvester (1884) showed that $f(a_1, a_2) = (a_1 - 1)(a_2 - 1) - 1$. Explicit solutions for $k = 3$ are known (Selmer and Beyer 1978, Rödseth 1978, Greenberg 1988). No closed-form solution is known for $k > 3$. For a big k, the problem of finding the Frobenius number is hard (**NP**-hard). The following theorem of Issai Schur (Figure 2.2) gives a more precise version of Theorem 2.33. We outline a short proof using elementary results in the theory of one complex variable.

Theorem 2.36. *Let $2 \le a_1 < \cdots < a_k$ be $k \ge 2$ positive integers whose gcd is 1. For $m \in \mathbb{N}$, let $a_m \in \mathbb{Z}_+$ be the number of ways that m can be expressed as $m = b_1 a_1 + \cdots + b_k a_k$ with nonnegative integers $b_1, \ldots, b_k$. Then*

Figure 2.2: Issai Schur (1875–1941). Credit: Wikipedia. Source: https://en.wikipedia.org/wiki/Issai_Schur.

$$\lim_{m \to \infty} \frac{a_m (k-1)! a_1 \ldots a_k}{m^{k-1}} = 1. \tag{2.2.3}$$

Proof. For $q \in \mathbb{N}$, consider the polynomial $s_q(x) := 1 - x^q$. The roots of this polynomial are all qth roots of unity:

$$x_{j,q} := e^{\frac{2\pi j \sqrt{-1}}{q}}, \quad j = 0, \ldots, q - 1.$$

Here $\sqrt{-1}$ stands for the *imaginary* complex number. Sometimes, we use the notation $\mathbf{i}$ to represent $\sqrt{-1}$. Note that $x_{0,q} = 1$. Let $p(x) = \prod_{i=1}^{k} s_{a_k}(x)$. Then the roots of $p(x)$ are the union of all a_ith roots of unity for $i = 1, \ldots, k$. Note that 1 is a root of p of multiplicity k. Since $a_1, \ldots, a_k$ are coprime, any other root ζ of $p(x)$ has multiplicity less than k. Let $\zeta_0 = 1, \ldots, \zeta_l$ be all the distinct roots of $p(x)$. Let $m_j \in \mathbb{N}$ be the multiplicity of the root ζ_j for $j = 0, \ldots, l$. So $k = m_0 > m_j$ for $j = 1, \ldots, l$. Thus $p(x) = (-1)^k \prod_{j=0}^{l} (x - \zeta_j)^{m_j}$. Consider the rational function $r(x) := \frac{1}{p(x)}$. Note that $r(0) = 1$ and $r(x)$ is analytic in the unit disk $|x| < 1$. Hence $r(x)$ has a power series expansion around $x = 0$ (Maclaurin expansion) with $a_i := \frac{r^{(i)}(0)}{i!}$ for $i = 0, 1, \ldots$. Use the geometric expansion $\frac{1}{1-t} = \sum_{i=0}^{\infty} t^i$ to deduce the identity

$$r(x) = \sum_{i=0}^{\infty} a_i x^i = \prod_{j=1}^{k} \sum_{n=0}^{\infty} x^{n a_j}.$$

Hence each a_m is a nonnegative integer, and for $m \in \mathbb{N}$, a_m is the number of ways that m can be represented as a nonnegative linear combination $b_1 a_1 + \cdots + b_k a_k$ with nonnegative integers $b_1, \ldots, b_k$. Hence m is expressible as such a linear combination if and only if $a_m > 0$. It is left to prove that the limit in (2.2.3) exists and is equal to 1. The partial fraction decomposition of $r(x)$ is of the form

$$r(x) = \sum_{j=0}^{l} \sum_{i=1}^{m_j} \frac{A_{ji}}{(1 - x\zeta_j^{-1})^i}. \tag{2.2.4}$$

Note that

$$A_{0k} = \lim_{x \to 1} \frac{(1 - x)^k}{p(x)} = \prod_{i=1}^{k} \lim_{x \to 1} \frac{(1 - x)}{1 - x^{a_i}} = \frac{1}{a_1 \dots a_k} > 0 \quad \text{(L'Hôpital rule)}.$$

It is left to show that the contribution to the Maclaurin coefficients of $r(x)$ by the terms given in (2.2.4) is dominated the Maclaurin coefficients of $\frac{A_{0k}}{(1-x)^k}$ for high enough power of x. Consider the Newton's binomial theorem (Appendix A.5) for $s \in \mathbb{N}$:

$$(1 - t)^{-s} = \sum_{i=0}^{\infty} \binom{i + s - 1}{s - 1} t^i.$$

Thus the contribution of $\frac{A_{0k}}{(1-x)^k}$ to the mth Maclaurin coefficient is of order $\frac{A_{0k}m^{k-1}}{(k-1)!}$. Use the fact that $A_{0k} > 0$, $k > m_j$, and $|\zeta_j| = 1$ for each $j > 0$ to induce that the contributions from other terms in (2.2.4) to the Maclaurin coefficients of $r(x)$ is of order at most m^{k-2}. Hence (2.2.3) holds.　　　　　　　　　　　　　　　　　　　　　　　　　　□

Historical note. The Frobenius problem boasts a rich history, featuring numerous applications, extensions, and connections to various research areas. A thorough overview of all facets of the problem is available in [2]. Determining the Frobenius number exactly is generally a challenging task. Brauer [10] successfully determined the Frobenius number for consecutive integers, whereas Roberts [35] expanded this achievement to numbers in arithmetic progression. Selmer (Figure 2.3) further gener-

Figure 2.3: Ernst Sejersted Selmer (1920–2006). Credit: Lisens: Universitetet i Bergen. Source: https://snl.no/Ernst_Sejersted_Selmer.

alized these findings to the determination of $f(a, ha + d, ha + 2d, \ldots, ha + nd)$ [38]. Only a few instances exist where the Frobenius number has been precisely determined for n variables. In the absence of exact results, research on the Frobenius problem has often been focused on sharpening bounds on the Frobenius number and algorithmic aspects.

? 2.3 Worked-out problems

1. Prove that a graph T is a tree if and only if there exists a unique path between every pair of distinct vertices in T.

   Solution. Assume that T is a tree. Then it is connected, and there is at least one path between any pair of vertices within the T Now consider two vertices u and v of T and suppose that there are two distinct paths between them. These two paths together form a circuit, indicating that T cannot be a tree.

2. Show that the complete bipartite graph $K_{m,m}$ is Hamiltonian if and only if $m > 1$.

   Solution. One direction is apparent as $K_{1,1}$ has no cycles, let alone a Hamiltonian cycle. Conversely, for $m > 1$, list the vertices alternately from the two partite sets. Consecutive vertices are adjacent, and the last is adjacent to the first, thus forming a Hamiltonian cycle.

3. Prove that a digraph G is strongly connected if and only if it possesses a closed spanning walk.

   Solution. Assuming that G is strongly connected, label the vertices as $v_1, \ldots, v_n$. Because of the strong connectivity of G, for each $1 \le i \le n - 1$, there exists a $v_i - v_{i+1}$ path, and there is also a $v_n - v_1$ path. Combining these n paths results in a closed spanning walk.
   Conversely, let us suppose there is a path between every pair of vertices in T. This implies that T is connected. If T has a cycle C and u and v are two vertices of C, then there are two distinct paths connecting u and b through C, which contradicts our initial assumption. Thus T is acyclic. We totally conclude that T is connected.

i 2.4 Problems

1. Prove Proposition 2.2.
2. Given an undirected graph G, prove that every $u - v$ walk in G contains a $u - v$ path.
3. Let G be an undirected graph. An edge $e \in V$ is called a *bridge* if the removal of e increases the number of connected components of G.
   (a) Let G be connected, and suppose that every spanning tree contains e. Prove that e is a bridge.
   (b) Let G be a tree. Prove that every edge in G is a bridge.

4. Prove Proposition 2.9.
5. Prove that an undirected graph is bipartite if and only if it is 2-colorable.
6. Prove that any undirected graph on n vertices with at least n edges must contain a cycle.
7. Prove Euler's formula for planar graphs; suppose we have a connected planar graph with and f faces of order n and size m. Then we have Euler's formula $n - m + f = 2$.

   (**Hint:** Use induction on m, the number of edges in the graph. For the inductive case, start with an arbitrary graph with m edges. Consider two cases: either G contains a cycle, or it does not.)
8. Given a connected planar graph G of order n and size m, prove that $m \leq 3n - 6$.
9. Verify that all complete bipartite directed graphs that are trees are stars.
10. Determine all trees that have Hamiltonian cycles.
11. Prove that every undirected graph with only vertices of even degree decomposes into cycles.
12. The graph shown below is the *Petersen* graph. Does it have a Hamilton cycle? Does it have a Hamilton path?

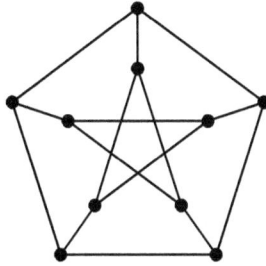

13. The Petersen graph is introduced in Problem 2.4-12. Find its diameter and girth.
14. Prove Proposition 2.22.
15. Show that a directed graph is an oriented graph if and only if it has no 2-cycles.
16. Prove that a digraph is acyclic if and only if it has no closed directed walks.
17. Given a connected digraph $G = (V, E)$, prove that the associated reduced graph $G_{\mathrm{rdc}} = (V_{\mathrm{rdc}}, E_{\mathrm{rdc}})$ is acyclic.
18. (a) Consider the following digraph $G = (V, E)$:

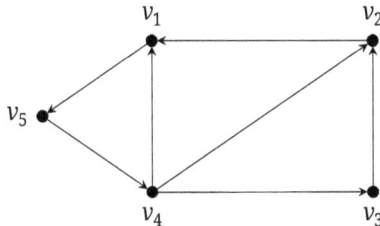

   Prove that there exists a positive integer N such that for all $m \geq N$ and $u, v \in V$, there exists a $u - v$ walk of length m.
   (b) Consider the following digraph $H = (V, E)$:

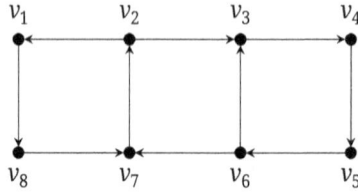

Prove that V can be partitioned into V_1 and V_2 such that (i) and (ii) hold simultaneously:

i. $E \cap (V_1 \times V_2) \neq \emptyset$ and $E \cap (V_2 \times V_1) \neq \emptyset$.

ii. $E \subset (V_1 \times V_2) \cup (V_2 \times V_1)$.

Moreover, prove that there exists a positive integer N such that for all $m \geq N$ and $u, v \in V_i$, there exists a walk from u to v of length $2m$ for each $i = 1, 2$.

19. Let $G = (V, E)$ be an undirected connected graph. Let $G_{max} = (V, E_{max})$ be the directed graph induced by converting each undirected edge uv to two directed edges (u, v) and (v, u) in the opposite directions. Then G_{max} has a cycle of length 2, $\{u, v, u\}$. Use Theorems 2.32 and 2.14 to prove that if $G = (V, E)$ is a connected undirected graph, then exactly one of the following conditions hold:

   (a) There exists a positive integer N such that for all $u, v \in V$ and $m \geq N$, there exists a walk W on G from u to v of length m.

   (b) G is bipartite, i. e., V is a union of two disjoint nonempty sets V_1 and V_2 such that $E \subset E_1 \times E_2$. Let $u, v \in V_i$ for some $i \in \{1, 2\}$. Then any walk from u to v has an even length. Furthermore, there exists a positive integer N such that for all $m \geq N$, there exists a walk W on G from u to v of length $2m$.

20. Prove Sylvester's formula; $f(a_1, a_2) = (a_1 - 1)(a_2 - 1) - 1$, where $f(a_1, a_2)$ stands for the Frobenius number of a_1 and a_2.

3 Random graphs

Outline. This chapter is structured into four sections. We commence with exploring two frequently employed models for random graphs: *Gilbert's $\mathcal{G}_{n,p}$ model* and *Erdös–Rényi's $\mathcal{G}_{n,m}$ model*. We briefly study their relationships and their implications for the analysis of graph connectivity. In Section 3.2, we bring up *threshold functions* and discuss graph properties through them. In the subsequent section, our exploration extends to the k-clique property of graphs, a tool for identifying the "dense" subgraphs within a given graph. In the final section, we delve into isolated vertices and connectivity, exploring threshold functions through the Poisson distribution.

3.1 Random graph models

A random graph is generated by initially having a set of n isolated vertices and subsequently adding successive edges between them at random. The focus of research in this field is on identifying the stage at which a specific graph property is likely to emerge. In other words, the initial motivation behind the study of random graphs was to comprehend the characteristics of "typical" graphs. This stands in contrast to the studies of "extremal" graphs, although it is worth noting that random graphs have, on occasion, exhibited properties more extreme than graphs derived through more constructive methods. Different random graph models result in different probability distributions on graphs. One extensively studied model is $\mathcal{G}_{n,p}$, introduced by Edgar Nelson Gilbert (Figure 3.3) [20], where each potential edge occurs independently with probability $p \in [0,1]$. A year after Gilbert's work, two famous combinatorialists Paul Erdös (Figure 3.1) and Alfréd Rényi (Figure 3.2) proposed a closely related model, as discussed in their seminal paper [12], in which they observed that many natural properties of undirected graphs exhibit a simple threshold function. We will illustrate both models starting with Gilbert's model for now.

Gilbert's $\mathcal{G}_{n,p}$ model. Let n be a positive integer, and let $p \in [0,1]$. Consider $\binom{n}{2}$ independent identically distributed Bernoulli random variables X_{ij} for $1 \leq i < j \leq n$ with $\mathbf{Pr}(X_{ij} = 1) = p$ and $\mathbf{Pr}(X_{ij} = 0) = 1 - p$. The random vector $\mathbf{X}_n := (X_{12}, \ldots, X_{(n-1)n}) \in \{0,1\}^{\binom{n}{2}}$ induces the undirected graph $G(\mathbf{X}_n) = ([n], E(\mathbf{X}_n))$ with the following edge assignments: for all $1 \leq i < j \leq n$,

$$(i,j) \in E(\mathbf{X}_n) \quad \text{if and only if} \quad X_{ij} = 1.$$

In other words, we start with an empty graph with vertex set $[n]$ and perform $\binom{n}{2}$ Bernoulli experiments including edges independently with probability p. We call such a randomly obtained graph a *binomial random* graph (or simply a *random* graph).

In an intuitive sense, the process begins with n isolated vertices. Independently, for every pair of vertices u and v:

https://doi.org/10.1515/9783111337388-003

–   An edge is added between u and v with probability p.
–   No edge is added with probability $1 - p$.

Parameterwise, n stands for the number of vertices, and p stands for the probability that an edge exists between any two vertices. In terms of randomness, each edge is added or not added independently, leading to a variety of possible graph structures. Notationwise, we may occasionally use $G_{n,p}$ to denote a random graph obtained through the $\mathcal{G}_{n,p}$ model.

Fixing n and p, let us denote by $\mathcal{G}_{n,p}$ the sample space consisting all random graphs $G = G(\mathbf{X}_n)$ induced by random vectors $\mathbf{X}_n$. We define a probability measure on $\mathcal{G}_{n,p}$ as follows:

$$\mathbf{Pr}_{n,p}(G) = p^{|E|}(1 - p)^{\binom{n}{2}-|E|},$$

where $|E|$ is the size (number of edges) of G, and $\binom{n}{2}$ is the number of all possible edges in an n-vertex graph. Notice that in the $\mathcal{G}_{n,p}$ model the expected number of edges is $\frac{np(n-1)}{2}$ approximately, because $\frac{n(n-1)}{2}$ is the total number of all possible edges and p, the prob-

Figure 3.1: Paul Erdös (1913–1996). Photo credit: Wikipedia. Source: https://en.wikipedia.org/wiki/Paul_Erd%C5%91s.

Figure 3.2: Alfréd Rényi (1921–1970). Photo credit: KritiK. Source: https://blog.kritik.co.uk/2018/07/conditional-probability-renyi-axioms.html.

Figure 3.3: Edgar Gilbert (1923–2013). Photo credit: Alchetron. Source: https://alchetron.com/Edgar-Gilbert.

ability of existence of each possible edge, is multiplied by this total to account for the randomness in the model.

We now highlight some important remarks about the $\mathcal{G}_{n,p}$ model:

– A random graph is not a graph in its own right; instead, it is a probability space with graphs as its elements.
– Each generated graph will likely be different due to the randomness involved.
– The connectivity properties of the graph will depend on p, that is, for higher values of p, the graph becomes more likely to be connected, forming a giant component that includes most vertices.

Example 3.1. We use Gilbert's $\mathcal{G}_{n,p}$ model for $n = 5$ and $p = 0.3$, that is, the graph starts with 5 isolated vertices. For each possible pair of vertices, a random decision is made with a chance of $p = 0.3$ to include an edge between them. We expect approximately $\frac{(5) \cdot (0.3) \cdot (4)}{2} = 3$ edges. Let us assume that we have obtained the following graph G of size 4:

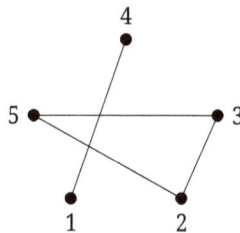

The probability of obtaining this specific graph among all 5-vertex graphs with probability parameter $p = 03$ is

$$\mathbf{Pr}_{5,0.3}(G) = 0.3^4(1 - 0.3)^{\binom{5}{2}-4} \approx 0.00095.$$

We now turn our attention to Erdös–Rényi's model.

Erdös–Rényi's $\mathcal{G}_{n,m}$ model. Let $\mathcal{G}_{n,m}$ be the family of all labeled graphs with vertex set $[n]$ and m edges. Clearly, $0 \leq m \leq \binom{n}{2}$. To each graph $G \in \mathcal{G}_{m,n}$, we assign the following probability:

$$\mathbf{Pr}(G) = \frac{1}{\binom{\binom{n}{2}}{m}}.$$

In other words, we begin with n isolated vertices, and in a random process, we keep adding edges to reach m edges ultimately in a way that all possible $\binom{\binom{n}{2}}{m}$ are equally likely. We denote such a randomly obtained graph by $G_{n,m} = ([n], E_{n,m})$ and refer to it as a *uniform random graph* (or simply a *random* graph.)

We now highlight some important remarks about Erdös–Rényi's $\mathcal{G}_{n,m}$ model:

- Each generated graph will likely be different due to the randomness involved.
- The expected average degree of a vertex is approximately $\frac{2m}{n}$.
- The connectivity properties of the graph will depend on the value of m, that is, for larger values of m, the graph becomes more likely to be connected.

Example 3.2. We use Erdös–Rényi's $\mathcal{G}_{n,m}$ model for $n = 5$ and $m = 4$. The graph starts with 5 isolated vertices. We choose 4 edges randomly from $\binom{5}{2} = 10$ possible edges and add them to the graph. The specific edges added will vary in different realizations of the model, but the overall structure will have 5 vertices and 4 edges. Let us assume that we have obtained the following graph G through the $\mathcal{G}_{5,4}$ model:

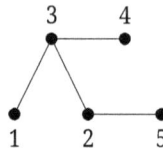

Then its degree sequence is $3, 2, 1, 1, 1$, and the expected average degree of a vertex is $\frac{(2) \cdot (4)}{5} = 1.6$. In addition, the probability of obtaining this specific graph among all 5-vertex graphs of size 4 is

$$\mathbf{Pr}(G) = \frac{1}{\binom{\binom{5}{2}}{4}} \approx 0.0048.$$

Comparison of random graph models; $\mathcal{G}_{n,p}$ versus $\mathcal{G}_{n,m}$. We first describe the differences between the two models. In Gilbert's model, edges are independently added between vertex pairs with fixed probability p, whereas in Erdös–Rényi's model, the graph has a constant number of edges m. In Gilbert's model, the density of the graph (the ratio of actual edges to possible edges) varies based on the chosen probability p, whereas in Erdôs–Rényi's model, the density is fixed since the number of edges m is constant. Gilbert's model produces a binomial distribution for the number of edges, whereas Erdös–Rényi's model produces a uniform distribution over all possible graphs with m edges. Gilbert's model offers flexibility in adjusting the graph's density through the probability parameter p, whereas Erdös–Rényi's model is less flexible in density control as m remains fixed. Having said this, Gilbert's model tends to be more prevalent in random graph theory. Despite these distinctions, the models share a close relationship, as explained in the following proposition.

Proposition 3.3. *A random graph G of size m obtained through Gilbert's $\mathcal{G}_{n,p}$ model is equally likely to be one of $\binom{\binom{n}{2}}{m}$ graphs of size m.*

Proof. Let G be a labeled graph of size m. Then we have

$$\{G_{n,p} : G_{n,p} = G\} \subset \{G_{n,p} : |E_{n,p}| = m\},$$

entailing

$$
\begin{aligned}
\mathbf{Pr}(G_{n,p} = G \,|\, |E_{n,p}| = m) &= \frac{\mathbf{Pr}(G_{n,p} = G, |E_{n,p}| = m)}{\mathbf{Pr}(|E_{n,p}| = m)} \\
&= \frac{\mathbf{Pr}(G_{n,p} = G)}{\mathbf{Pr}(|E_{n,p}| = m)} \\
&= \frac{p^m (1-p)^{\binom{n}{2}-m}}{\binom{\binom{n}{2}}{m} p^m (1-p)^{\binom{n}{2}-m}} \\
&= \frac{1}{\binom{\binom{n}{2}}{m}},
\end{aligned}
$$

as claimed. □

After all, which model is preferred? There is no definitive answer to the question of which model is better than the canonical model for a random graph. However, in the context of this question, it seems natural to consider $G_{n,m}$ since there is an evident extremal condition concerning edges. Specifically, we need at least $n-1$ edges before there is any possibility that the graph is connected. Therefore it is reasonable to inquire about the number of edges required before there is a strong likelihood that $G_{n,m}$ is connected. Certainly, conditioning on m edges being present in $G_{n,p}$ yields $G_{n,m}$. We can perceive $G_{n,p}$ as a mixture of distributions $G_{n,m}$ across values of m from 0 to $\binom{n}{2}$. Some graph properties are evidently poorly focused under this mixture, such as the property of the graph having an odd size. However, for a monotone-increasing graph property, verifying that an event holds w. h. p for $G_{n,p}$ given it holds w. h. p for $G_{n,m}$, and vice versa, is usually straightforward. For a comprehensive discussion on probabilistic methods, particularly random graphs, the interested reader is referred to the masterful work by Alon and Spencer [4]. Also, a collection of lectures on the applications of probabilistic ideas to problems in combinatorics with informal language may be found in [40].

3.2 Thresholds

We begin this section with the coupling technique, a useful tool in random graph theory to prove theorems about graph properties and phase transitions.

Coupling technique. Coupling is a method for comparing two probability distributions by constructing a joint distribution that links them together. We first restrict the discussion to random graphs $G_{n,p}$ obtained through Gilbert's $\mathcal{G}_{p,n}$ model. Suppose $0 < p_1 < p$. Find $p_2 \in (0,1)$ such that $p = p_1 + p_2 - p_1 p_2$ or, in other words, $1 - p = (1-p_1)(1-p_2)$. Therefore $G_{n,p}$ does not contain an edge if and only if it is not included in G_{n,p_1} or G_{n,p_2}. Therefore

$$G_{n,p} = G_{n,p_1} \cup G_{n,p_2}.$$

Thus the notation $G_{n,p_1} \subset G_{n,p}$ means that the two graphs are *coupled* so that $G_{n,p}$ is obtained from G_{n,p_1} by attaching it to G_{n,p_2} and replacing the repeated edges by a single one.

We can also discuss coupling in the context of Erdös–Rényi's model. Let us couple random graphs G_{n,m_1} and G_{n,m_2} obtained through $\mathcal{G}_{n,m_1}$ and $\mathcal{G}_{n,m_2}$, respectively, where $m_2 > m_1$. Let us couple G_{n,m_1} and G_{n,m_2} through

$$G_{n,m_1} = G_{n,m_2} \cup G,$$

where G is the random graph on the vertex set $[n]$ that has $m = m_2 - m_1$ edges chosen uniformly at random from $\binom{n}{2} \setminus E_{n,m_1}$. In other words, we first pick m_1 edges that belong to G_{n,m_1} and then from the remaining edges pick $m = m_2 - m_1$ edges that belong to G_{n,m_2} but not to G_{n,m_1}.

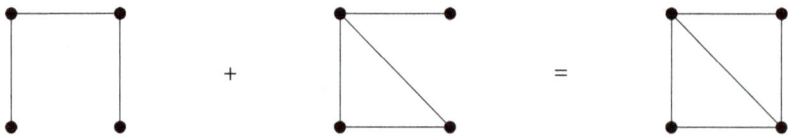

Threshold functions. We begin with the monotone properties of graphs. A *graph property* is defined as a characteristic that remains unchanged under all possible isomorphisms of a graph. In other words, it is a property of the graph itself, not of a specific drawing or representation of the graph. Let $\mathcal{A}$ be a graph property, that is, if a graph G belongs to $\mathcal{A}$ and $G \cong H$, then $H \in \mathcal{A}$. For example, having a cycle is a graph property. Now let $\mathcal{A}$ be a property of an undirected graph. We say that $\mathcal{A}$ is *monotone-increasing* if $G \in \mathcal{A}$ implies $G + e \in \mathcal{A}$, where e is an edge, that is, adding an edge e to G does not compromise the property. We say that $\mathcal{A}$ is *monotone-decreasing* if $G \in \mathcal{A}$ implies $G - e \in \mathcal{A}$. In other words, removing an edge does not compromise the property. We say that $\mathcal{A}$ is *monotone* if $\mathcal{A}$ is either monotone-increasing or monotone-decreasing. A monotone-increasing property $\mathcal{A}$ is *nontrivial* if the empty graph $K_n^c \notin \mathcal{A}$ and $K_n \in \mathcal{A}$.

Clearly, connectivity is a monotone-increasing property, whereas being planar is a monotone-decreasing property. Furthermore, not all graph properties are monotone (Why? See Problem 3.6-3(a).)

Now let $\mathcal{A}$ be a graph property. Let $\mathcal{A}_n \subset \mathcal{G}_{n,p}$ be the set of all n-vertex graphs with property $\mathcal{A}$. Define $f_n(p) := \mathbf{Pr}_{n,p}(\mathcal{A}_n)$. We say that $\mathcal{A}$ is a *monotone property in p* if $f_n(p)$ is a monotone function. Wee may argue that $\mathcal{A}$ is a monotone property in p if and only if $\mathcal{A}$ is a monotone property in the above sense. Using the coupling technique, we may argue that if $\mathcal{A}$ is a monotone-increasing property and $p_1 < p_2$, then $f_n(p_1) < f_n(p_2)$. (See Problem 3.6-4.)

We now use the concept of monotone properties to define threshold functions. A function $p : \mathbb{N} \to [0,1]$ is called a *threshold* function for the monotone-increasing property $\mathcal{A}$ in the random graph $G_{n,p}$ if for every function $r : \mathbb{N} \to [0,1]$, the following two conditions hold:

(a) $\lim_{n\to\infty} \frac{r(n)}{p(n)} = 0$ implies that $\lim_{n\to\infty} \mathbf{Pr}(G \in \mathcal{G}_{n,r(n)}$ has $\mathcal{A}) = 0.$

(b) $\lim_{n\to\infty} \frac{r(n)}{p(n)} = \infty$ implies that $\lim_{n\to\infty} \mathbf{Pr}(G \in \mathcal{G}_{n,r(n)}$ has $\mathcal{A}) = 1.$

In other words, p is a threshold function for $\mathcal{A}$ in $G_{n,p}$ if
(a) For every r with $r \ll p$, then $G_{n,r(n)}$, w. h. p, does not have the property $\mathcal{A}$.
(b) For every r with $p \ll r$, then $G_{n,r(n)}$, w. h. p, has the property $\mathcal{A}$.

In this case, we say that a *phase transition* occurs. This phenomenon is akin to the abrupt phase transitions observed in physics when temperature or pressure increases. Examples of this include the abrupt appearance of cycles in $G_{n,p}$ as p approaches $\frac{1}{n}$. (Worked-out problem 3.5-1.)

Similarly, a threshold function may be defined for $G_{n,m}$: A function $m : \mathbb{N} \to [0,1]$ is called a *threshold* function for the monotone-increasing property $\mathcal{A}$ in the random graph $G_{n,m}$ if for every function $r : \mathbb{N} \to [0,1]$, the following two conditions hold:
(a) $\lim_{n\to\infty} \frac{r(n)}{m(n)} = 0$ implies that $\lim_{n\to\infty} \mathbf{Pr}(G \in \mathcal{G}_{n,r(n)}$ has $\mathcal{A}) = 0.$
(b) $\lim_{n\to\infty} \frac{r(n)}{m(n)} = \infty$ implies that $\lim_{n\to\infty} \mathbf{Pr}(G \in \mathcal{G}_{n,r(n)}$ has $\mathcal{A}) = 1.$

A significant portion of the random graph theory focuses on identifying thresholds for various properties. These properties include containing a path or cycle of a specified length, having a copy of a given graph, exhibiting connectivity, or being Hamiltonian, among others. We list a couple of threshold functions in the random graph $G_{n,p}$:

Property	Threshold
Contains a path of length k	$p(n) = n^{-\frac{k+1}{k}}$,
Is not planar	$p(n) = \dfrac{1}{n}$,
Contains a Hamiltonian path	$p(n) = \dfrac{\log n}{n}$,
Contains a cycle	$p(n) = \dfrac{1}{n}$,
Is connected	$p(n) = \dfrac{\log n}{n}$,
Contains a clique on k points	$p(n) = n^{-\frac{2}{k-1}}$.

Also, a list of a couple of threshold functions in the random graph $G_{n,m}$:

Property	Threshold
Is connected	$m(n) = n \log n$,
k-Colorability	$m(n) = \dfrac{(k-2)n}{2}$,
Contains a giant component	$m(n) = \left(\dfrac{1}{2} + \sqrt{\dfrac{\log n}{n}}\right)^n$.

Notice that, for example, for the property of being connected, $p(n) = \frac{\log n}{n}$ means that a random graph with n vertices is likely to be connected if it has the probability parameter $p \geq p(n)$.

We now emphasize two crucial observations regarding threshold functions:

- The thresholds described earlier are not unique, as any function differing from p (or m) by a constant factor also serves as a threshold for $\mathcal{A}$.
- In the random graphs $G_{n,p}$ and $G_{n,m}$, assuming that $m \approx p\binom{n}{2}$, the two models often produce graphs with similar global properties that depend on overall edge density or connectivity. (This includes properties like connectivity, the size of the largest component, and the diameter of the graph.) For example, as we addressed above (and will prove later in this chapter), $p(n) = \frac{\log n}{n}$ is the threshold for connectedness. We will see that $m(n) = \binom{n}{2}\frac{\log n}{n} \approx \frac{1}{2}n \log n$ is a threshold for connectedness for $G_{n,m}$.
- Every nontrivial monotone graph property possesses a threshold.

The latter statement, attributed to Bollobás and Thomason [9], stands as a fundamental theorem in random graph theory.

Evolution of Random Graphs: From Inception to Present. The inception of a random graph model can be traced back to 1938 when Helen Hall Jennings and Jacob Moreno introduced a "chance sociogram," essentially a directed Erdös–Rényi model. Their aim was to study and compare the proportion of reciprocated links in their network data with the random model [26]. In another instance, Solomonoff and Rapoport employed chance sociogram to study social networks, similar to the Erdös–Rényi model. Edgar Nelson Gilbert introduced the $\mathcal{G}_{n,p}$ model, emphasizing connectivity properties. Erdös and Rényi pioneered the field by formally defining two fundamental models $\mathcal{G}_{n,m}$ and $\mathcal{G}_{n,p}$ understanding threshold phenomena, connectivity thresholds, and the emergence of giant connected components. Their work laid the foundation for probabilistic graph theory, exploring various properties of random graphs. Béla Bollobás (Figure 3.4) made substantial contributions to random graph theory and is renowned for his influential book "Random Graphs" [8]. In the 1960s, phase transitions in random graphs were identified, where properties abruptly change with varying parameters. Since the 1960s, various models emerged, including small-world, scale-free, and exponential random graphs, capturing diverse network structures. On the other hand, from a technical perspective, the evolution of random graphs studies how $G_{n,p}$ and $G_{n,m}$ evolve as p increases from 0 to 1 and m increases from 0 to $\binom{n}{2}$.

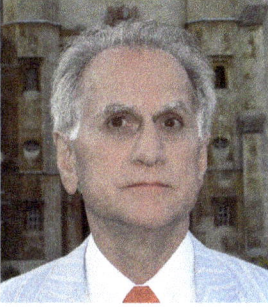

Figure 3.4: Béla Bollobás (1943 (age 81)). Photo credit: Wikipedia. Source: https://en.wikipedia.org/wiki/B%C3%A9la_Bollob%C3%A1s.

In the next two sections, we first explore the threshold function of k-clique property ($p(n) = n^{-\frac{2}{k-1}}$) and then the threshold function of the presence of isolated vertices that is equivalent to graph being connected ($p(n) = \frac{\log n}{n}$). In these scenarios, more precise results beyond the threshold function can be derived by employing the Poisson distribution.

3.3 *k*-clique property

For a fixed constant $k \geq 3$, the k-clique problem is the task of determining whether a graph of order n contains a complete subgraph of order k. We say that a graph possesses the *k-clique property* if it contains at least one k-clique. We now present the problem formally. Let $G = ([n], E)$ be an undirected graph on n vertices. Let $S \subset [n]$ be a k-clique of G. Define the Bernoulli random variable $X_S : \mathcal{G}_{n,p} \rightarrow \{0, 1\}$ as follows:

"For $G \in \mathcal{G}_{n,p}, X_S(G) = 1$ if and only if G contains a clique on the set S."

In other words, $X_S(G) = 1$ if and only if $(i,j) \in E(G)$ for all $i,j \in S$.

We claim that $\mathbf{Pr}_{n,p}(X_S = 1) = p^{\binom{|S|}{2}}$. Indeed, to have a clique on S, we must have all edges (i,j) for each pair $i,j \in S$, that is, each random variable X_{ij} is equal to 1. Fix $k \geq 2$ and let $n \geq k$. Let $X_n = \sum_{S \subset [n], |S|=k} X_S$. Then $X_n : \mathcal{G}_{n,p} \to \mathbb{Z}_+$ is a random variable such that $X_n(G)$ counts the number of k-cliques in a graph G on n vertices. Since the number of all possible k-cliques in a graph on n vertices is $\binom{n}{k}$, i.e., the number of all distinct choices of the sets of k elements in $[n]$, it follows that

$$E_p(X_n) = \sum_{S \subset [n], |S|=k} E_p(X_S) = \binom{n}{k} p^{\binom{k}{2}}$$

$$= \frac{n(n-1)\ldots(n-k+1)}{1 \cdot 2 \cdot \ldots \cdot k} p^{\frac{k(k-1)}{2}} = (np^{\frac{k-1}{2}})^k \frac{1}{k!} \prod_{j=0}^{k-1}\left(1 - \frac{j}{n}\right). \tag{3.3.1}$$

Here we emphasize the fact that our expectation on $\mathcal{G}_{n,p}$ depends on p. Hence for sequences $r(n) \in [0,1], n \in \mathbb{N}$, we have the following implication:

$$\lim_{n\to\infty} n^{\frac{2}{k-1}} r(n) = a \in [0,\infty] \Rightarrow \lim_{n\to\infty} E_{r(n)}(X_n) = \frac{a^{\binom{k}{2}}}{k!}. \tag{3.3.2}$$

In particular, if $\lim_{n\to\infty} n^{\frac{2}{k-1}} r(n) = 0$ or ∞, then the expected value of the number of k-cliques tends to zero or infinity, respectively. Thus the threshold function $p(n) = n^{-\frac{2}{k-1}}$ essentially gives the following rough information. If $\lim_{n\to\infty} n^{\frac{2}{k-1}} r(n) = 0$ or ∞, then the probability of having no k-clique is approaching to one or zero, respectively, as $n \to \infty$.

We first show that property (a) of the threshold function $p(n) = n^{-\frac{2}{k-1}}$ follows simply from the fact that the expected number of cliques tends to 0. For that, we need the following lemma.

Lemma 3.4. *Let Ω be a countable sample space with probability measure. Let $X : \Omega \in \{0\} \cup [1,\infty)$ be a random variable. Then $\mathbf{Pr}(X \geq 1) \leq E(X)$.*

Proof. It is immediate from Theorem 1.8 for $a = 1$. $\square$

We proceed with our discussion concerning verifying property (a). Since $X_n \in \mathbb{Z}_+$, we deduce that

$$\lim_{n\to\infty} n^{\frac{2}{k-1}} r(n) = 0 \Rightarrow 1 \geq \mathbf{Pr}_{n,r(n)}(X_0) = 1 - \mathbf{Pr}_{n,r(n)}(X_n \geq 1) \geq 1 - E_{n,r(n)}(X_n) \to 1,$$

and property (a) follows.

To show property (b), we use Chebyshev's inequality. Clearly, $X_n = 0$ satisfies the inequality $|X_n - E_{n,p}| \geq |E_{n,p}(X_n)| = E_{n,p}(X_n)$ for all $n \geq k$ and $p \in [0,1]$. Chebyshev's inequality (Theorem 1.12) yields $\mathbf{Pr}(|X_n - E_{n,p}| \geq |E_{n,p}(X_n)|) \leq \frac{\mathrm{Var}_{n,p}(X_n)}{E_{n,p}(X_n)^2}$. Thus we need to show that

$$\lim_{n\to\infty} n^{\frac{2}{k-1}} r(n) = \infty \Rightarrow \lim_{n\to\infty} \frac{\text{Var}_{n,r(n)}(X_n)}{\text{E}_{n,r(n)}(X_n)^2} = 0. \tag{3.3.3}$$

Notice that $X_n = \sum_{S\subset[n],|S|=k} X_S$, where each X_S is Bernoulli. If for any $S \neq T$, X_S and X_T are independent, then (1.3.4) yields that

$$0 \le \text{Var}_{n,p}(X_n) = \sum_{S\subset[n],|S|=k} \text{Var}_{n,p}(X_S) \le \sum_{S\subset[n],|S|=k} \text{E}_{n,p}(X_S) = \text{E}_{n,p}(X_n),$$

and (3.3.3) easily follows. However X_S and X_T are independent if $|S \cap T| \le 1$. Hence we need to use the second equality of (1.3.3), rather than (1.3.4). Note that since S and T are sets of a fixed size k and $n \to \infty$, it follows that for most of pairs $S, T \subset [n]$ of cardinality k, $S \cap T = \emptyset$. Hence most of the pairs X_S, X_T are independent Bernoulli variables for large n, that is, in (1.3.3), $\text{Cov}(X_S, X_T) = 0$ for $|S \cap T| \le 1$. Therefore (3.3.3) holds.

We now prove (3.3.3) for $k = 4$. Since the k-clique property is increasing in p. (The larger the probability, it is more likely you have more edges in the graph and hence greater chances to have a k-clique.) It suffices to show property (b) under the condition

$$r(n) \le \frac{\log n}{n^{\frac{2}{3}}} \quad \text{for } n \in \mathbb{N}, \quad \text{and} \quad \lim n^{\frac{2}{3}} r(n) = \infty. \tag{3.3.4}$$

First, observe that X_S and X_T are independent if $S \cap T = \emptyset$ since S and T are disjoint sets. Assume next that $|S \cap T| = 1$. Then the set of edges on a complete graph on S and T are disjoint sets. Hence X_S and X_T are independent. In both cases, $\text{Cov}(X_S, X_T) = 0$.

Assume next that $|S \cap T| \in \{2, 3, 4\}$. Since X_S and X_T are Bernoulli, $\text{E}(X_S), \text{E}(X_T) \ge 0$, and $X_S X_T$ is Bernoulli. Hence

$$\text{Cov}_{n,p}(X_S, X_T) = \text{E}_{n,p}(X_S X_T) - \text{E}_{n,p}(X_S)\text{E}_{n,p}(X_T)$$
$$\le \text{E}_{n,p}(X_S X_T) = \mathbf{Pr}_{n,p}(X_S X_T = 1) = \mathbf{Pr}_{n,p}(X_S = 1, X_T = 1).$$

Consider the case $|S \cap T| = 2$. Then $X_S = X_T = 1$ means that we have 12 edges in the complete graph on S and T, of which one edge is a joint edge. Thus we have 11 edges. Hence $\mathbf{Pr}_{n,p}(X_S = 1, X_T = 1) = p^{11}$. The number of such S, T is found as follows. First, choose two joint vertices in $S \cap T$. The number of such vertices is $\binom{n}{2}$. Then choose the remaining two vertices in S. The number of such choices is $\binom{n-2}{2}$. Then choose the other two vertices in T. Their number is $\binom{n-4}{2}$. Thus the total number of 4-sets S, T satisfying the condition $|S \cap T| = 2$ is $\binom{n}{2}\binom{n-2}{2}\binom{n-4}{2} \le n^6$. Hence

$$\sum_{S,T\subset[n],|S|=|T|=4,|S\cap T|=2} \text{Cov}_{n,r(n)}(X_S, X_T) \le n^6 \left(\frac{\log n}{n^{\frac{2}{3}}}\right)^{11} = \frac{(\log n)^{11}}{n^{\frac{4}{3}}} \to 0.$$

Consider the case $|S \cap T| = 3$. Then the complete graphs on $S \cap T$ have three common edges. So the number of total edges in the complete graphs on S and T is $12 - 3 = 9$. Hence $\mathbf{Pr}_{n,p}(X_S = 1, X_T = 1) = p^9$. The number of such S, T is found as follows. First, choose three joint vertices in $S \cap T$. The number of such vertices is $\binom{n}{3}$. Then choose the remaining vertices in S. The number of such choices is $n - 3$. Then choose the remaining vertices in T. Their number is $n - 4$. Thus the total number of 4-sets S, T satisfying the condition $|S \cap T| = 3$ is $\binom{n}{3}(n - 3)(n - 4) \leq n^5$. Hence

$$\sum_{S,T \subset [n],|S|=|T|=4,|S\cap T|=3} \mathrm{Cov}_{n,r(n)}(X_S, X_T) \leq n^5 \left(\frac{\log n}{n^{\frac{2}{3}}} \right)^9 = \frac{(\log n)^9}{n} \to 0.$$

Thus

$$0 \leq \frac{\mathrm{Var}_{n,r(n)}(X_n)}{\mathrm{E}_{n,r(n)}(X_n)^2}$$

$$= \frac{1}{\mathrm{E}_{n,r(n)}(X_n)^2} \left(\sum_{S \subset [n],|S|=4} \mathrm{Var}_{n,r(n)}(X_S) + \sum_{S,T \subset [n],|S|=|T|=4,|S\cap T| \in \{2,3\}} \mathrm{Cov}_{n,r(n)}(X_S, X_T) \right)$$

$$\leq \frac{\mathrm{E}_{n,r(n)}(X_n) + \frac{(\log n)^{11}}{n^{\frac{4}{3}}} + \frac{(\log n)^9}{n}}{\mathrm{E}_{n,r(n)}(X_n)^2} \to 0.$$

This shows (3.3.3) for $k = 4$ and completes the proof that $p(n) = n^{-\frac{2}{k-1}}$ is a threshold function for a k-clique in the case $k = 4$.

In fact, a stronger result holds.

Theorem 3.5. *Assume the equality in the first part of (3.3.2) with $a \in (0, \infty)$. Then X_n converges in probability to Poisson distribution $\mathrm{Pu}(b)$ with $b = \frac{a^{\binom{k}{2}}}{k!}$, that is, the probability that a random graph will have exactly j k-cliques is $e^{-b} \frac{b^j}{j!}$ for all $j \in \mathbb{Z}_+$.*

So as $a \to 0$, we obtain that with probability 1 a random graph has no k-clique. As $a \to \infty$, with probability 1 the random graph contains j k-cliques for every $j \geq 1$.

To prove the theorem, we need to use Theorem 1.23, that is, we need to show that

$$\lim_{n \to \infty} \mathrm{E}_{n,r(n)} \left(\sum_{S_1,...,S_j \subset [n],|S_1|=\cdots=|S_j|=k, S_i \neq S_l \text{ for } i \neq l} X_{S_1} \ldots X_{S_j} \right) = b^j \quad \text{for each } j = 2, \ldots .$$

Note that if $S_i \cap S_l = \emptyset$ for $i \neq l$, then $\mathrm{E}(X_{S_1} \ldots X_{S_j}) = \mathrm{E}(X_{S_1}) \ldots \mathrm{E}(X_{S_j})$. Hence we have to show that the contribution of other terms is negligible. To show that, we can do similar computations as we did for the case $k = 2$ for $\mathrm{Var}(X_n)$, which is equivalent to consider the case $j = 2$.

3.4 Isolated vertices and connectivity

Theorem 3.6. *Let $c \in \mathbb{R}$ and $p(n) = \frac{\log n + c}{n}$, $n \in \mathbb{N}$. Then $\lim_{n \to \infty} \mathrm{Pr}(G \in \mathcal{G}_{n,p(n)}$ does not have an isolated vertex) $= e^{-e^{-c}}$.*

Proof. Let X_i be the Bernoulli variable corresponding to the event A_i: the vertex i is isolated. Then $\textbf{Pr}_{n,p}(X_i = 1) = (1 - p)^{n-1}$. Note that

$$(1 - p(n))^{n-1} = e^{(n-1)\log(1-p(n))} = e^{-p(n)(n-1)+O((n-1)p(n)^2)}$$

$$= e^{-\log n - c + O(\frac{(\log n)^2}{n})} = \frac{e^{-c}}{n}\left(1 + O\left(\frac{(\log n)^2}{n}\right)\right).$$

Let $X_n = \sum_{i=1} X_i$ be the random variable on $\mathcal{G}_{n,p}$ such that $X_n(G)$ is the number of isolated vertices in G. Then $\mathrm{E}_{n,p(n)}(X_n) = (1 + O(\frac{(\log n)^2}{n}))e^{-c} \to e^{-c}$. We claim that X_n converges in probability to the Poisson distribution $\mathrm{Pu}(b)$ with $b = e^{-c}$. We use Theorem 1.23. Note that $\textbf{Pr}_{n,p}(X_{i_1} = 1, \ldots, X_{i_k} = 1) = (1 - p)^{(n-1)k-\binom{k}{2}}$, when $1 \le i_1 < \cdots < i_k \le n$. Indeed, this is the event that the given k vertices are isolated. Consider the complete graph K_n on n vertices. The degree of each vertex is $n - 1$. So there are exactly $(n - 1)k$ edges coming out of these k vertices. $\binom{k}{2}$ edges are common to both vertices in these groups. Hence the total number of (distinct) edges that are connected to these k vertices is $(n - 1)k - \binom{k}{2}$, implying $\mathrm{E}_{n,p(n)}(X_{i_1} = 1, \ldots, X_{i_k} = 1) \approx (\frac{b}{n})^k$. There are exactly $\binom{n}{k}$ of choices of k distinct vertices out of n. Therefore

$$\lim_{n\to\infty} \mathrm{E}_{n,p(n)}\left(\sum_{1\le i_1 < \cdots < i_k \le n} X_{i_1} \ldots X_{i_k}\right) = \frac{b^k}{k!} \quad \text{for } k = 1, 2, \ldots .$$

Note that the event $G \in \mathcal{G}_{n,p(n)}$ does not have an isolated vertex is equivalent to $X_n = 0$. As $\textbf{Pr}(\mathrm{Pu}(b) = 0) = e^{-b}$, we deduce the theorem. $\square$

It can be shown that under the assumptions of Theorem 3.6, as $n \to \infty$, the probability that G does not have an isolated vertex is equal to the probability that G is connected [12]. In other words, under the assumptions of Theorem 3.6, $\lim_{n\to\infty} \textbf{Pr}(G \in \mathcal{G}_{n,p(n)}$ is connected$) = e^{-e^{-c}}$. Therefore $p(n) = \frac{\log n}{n}$ is the threshold function for the connectivity property.

Further exploration: Chromatic Numbers of Random Graphs $G_{n,p}$. Significant progress in random graph theory has been achieved through the study of chromatic number concentration and the development of the probabilistic method. Noga Alon (Figure 3.5), along with Michael Krivelevich, proved that in the random graph $G_{n,p}$, the chromatic number is highly concentrated within a narrow range, deepening our understanding of its predictable behavior in large random graphs [3]. This research is closely connected to Alon's co-authored book, *The Probabilistic Method* [4], which employs probabilistic techniques to demonstrate the existence of combinatorial structures. These methods serve as powerful tools for analyzing and establishing properties of random graphs, underscoring their practical and theoretical significance in combinatorics.

Figure 3.5: Noga Alon (1956 (age 68)). Photo credit: Wikipedia. Source: https://en.wikipedia.org/wiki/Noga_Alon.

3.5 Worked-out problems

1. Prove that $p(n) = \frac{1}{n}$ is a threshold function for the existence of cycles in the random graph $G_{n,p}$.

Solution. We first verify the first condition of $p(n)$ being threshold. Let $r(n)$ be a function for which $\lim_{n\to\infty} \frac{r(n)}{p(n)} = 0$. In other words, $\lim_{n\to\infty} nr(n) = 0$. Sampling the graph G_n from $\mathcal{G}_{n,r(n)}$, we want to show that G_n has no cycle. Define the random variable X_n that counts the number of cycles in G_n. Applying Markov's inequality, we have

$$\mathbf{Pr}(G_n \text{ has a cycle}) = \mathbf{Pr}(X_n \geq 1) \leq \mathrm{E}(X_n). \tag{3.5.1}$$

Using (3.5.1), it suffices to show that $\lim_{n\to} \mathrm{E}(X_n) = 0$. Now let $\mathcal{C}_k$ be the set of all k-cycles in G_n up to rotation and orientation of the cycles, and let $\mathcal{C} = \cup_{k\geq 3}\mathcal{C}_k$. We now find $|\mathcal{C}_k|$, that is, we count the number of k-cycles. It would be equal to the number of ordered sets of size k from $[n]$, which is $\binom{n}{k}k!$. We encounter double counting in two ways: initially, for each starting position on the cycle (multiplied by k), and secondly, in choosing the direction of each cycle (multiplied by 2). Consequently, the total overcount is $2k$ times. Hence

$$|\mathcal{C}_k| = \frac{\binom{n}{k}k!}{2k} = \frac{\binom{n}{k}(k-1)!}{2}. \tag{3.5.2}$$

On the other hand, $\binom{n}{k}(k-1)! = \frac{n(n-1)\cdot(n-k+1)}{k} \leq \frac{n^k}{k}$. So we conclude that $|\mathcal{C}_k| \leq n^k$. For every $C \in \mathcal{C}_k$, the probability that a cycle occurs on C is $r(n)^k$ as we need each of the k independent edges that form the cycle to be present in our random graph. Therefore

$$E(X_n) = \sum_{k \geq 3} \binom{n}{k} \frac{(k-1)!}{2} r(n)^k \leq \sum_{k \geq 3} n^k r(n)^k = \frac{n^3 r(n)^3}{1 - nr(n)} \to 0$$

as $\lim_{n \to \infty} nr(n) = 0$. So the first condition of $p(n) = \frac{1}{n}$ being threshold is the case. We now verify the second condition of $p(n) = \frac{1}{n}$ being threshold. Let $r(n)$ be a function such that $\lim_{n \to \infty} \frac{r(n)}{p(n)} = \infty$. In other words, $\lim_{n \to \infty} nr(n) = \infty$. We define $q(n) = \frac{3}{n}$. Then clearly $q(n) \ll r(n)$. It suffices to show that $G_{n,q(n)}$ has a cycle w. h. p. Invoking Problem 3.6-6, it suffices to show that $G_{n,q(n)}$ has at least n edges. Sampling G_n from $G_{n,q(n)}$, we define the random variable X_n for the number of edges in G_n. Clearly, $X_n = \sum_{1 \leq i \leq j \leq n} 1_{(i,j) \in E(G_n)}$ is the sum of $\binom{n}{2}$ independent random variables each of which is 1 with probability $q(n) = \frac{3}{n}$ and 0 with probability $1 - q(n)$. Thus X_n has the binomial distribution, $X_n \sim Y(\binom{n}{2}, q)$, whose expectation and variance are

$$E(X_n) = \binom{n}{2}q \quad \text{and} \quad Var(X_n) = \binom{n}{2}q(1-q).$$

Therefore $E(X_n) = \binom{n}{2}\frac{3}{n} = \frac{3n}{2}(1 - \frac{1}{n})$. Now for $n > 8$, $|\frac{3n}{2}(1 - \frac{1}{n}) - b| < \frac{n}{3}$ implies $b > n$. Thus, to show that G_n has at least n edges, it suffices to show that $|E(X_n) - X_n| < \frac{n}{3}$, where Chebyshev's inequality is applicable for $a = \frac{n}{3}$:

$$Pr\left(|E(X_n) - X_n| \geq \frac{n}{3}\right) \leq \frac{Var(X_n)}{(\frac{n}{3})^2} = \frac{27}{2n}\left(1 - \frac{1}{n}\right)\left(1 - \frac{3}{n}\right) \to 0.$$

Therefore G_n has at least n edges, implying that G_n has a cycle w. h. p for $nr(n) \to \infty$.

3.6 Problems

1. Prove that in $G_{n,p}$ the probability of a vertex having degree k is $\binom{n}{k}p^k(1-p)^{n-k}$.
2. Let $0 < p(n) < 1$ be a function. Let A_1 be a graph property that holds for almost all graphs in $G_{n,p}$, and let A_2 be another graph property that holds for almost all graphs in $G_{n,p}$. Prove that
   (a) $A_1 \cap A_2$ is also a graph property.
   (b) the graph property $A_1 \cap A_2$ holds for almost all graphs in $G_{n,p}$.
3. Show that
   (a) not all graph properties are monotone.
   (b) a graph property in monotone-increasing if and only if its complement is monotone-decreasing.
   (c) if A is the property of being an Eulerian graph, then A is not monotone-increasing.
4. Let A be a monotone-increasing property, and let $0 < p_1 < p_2 < 1$. Use the coupling technique to prove that $f_n(p_1) < f_n(p_2)$, where the functions f_n are defined in Section 3.2.

5. (a) Construct a graph $G_{n,\frac{d}{n}}$ with $n = 50$ and $d = 2, 3$.
    (b) Determine the number of 3-cycles in each graph.
    (c) Repeat the experiment with $n = 500$.
    (d) Let $\mathcal{A}$ be the graph property of having a 3-cycle. Fix $p \in (0, 1]$. Prove that

$$\lim_{n \to \infty} \mathbf{Pr}(G_{n,p} \in \mathcal{A}) = 1.$$

6. Prove that the expected number of cycles of length k in $G \in \mathcal{G}_{n,p}$ is

$$\frac{n(n-1)(n-2)\cdots(n-k+1)}{2k} \cdot p^k.$$

7. Let $p \geq \sqrt{\frac{\ln n}{n-2}}$. Prove that the diameter of $G_{n,p}$ is at most 2 with high probability.

8. Prove that $p(n) = (\frac{1}{n})^{\frac{k}{k-1}}$ is the threshold function for the property of containing any tree on k vertices in $\mathcal{G}_{n,p}$.

4 Matrices and graphs

Outline. This chapter commences with the notion of adjacency matrices for graphs (undirected and directed graphs). By utilizing *adjacency matrices* we explore various graph characteristics, including vertex degrees, the enumeration of walks, the existence of odd cycles, and analysis of connectivity in graphs. Following that, we introduce block matrices and examine their relation to *strong connectivity* in digraphs. We finally discuss reduced graphs and reducible matrices. The reader is referred to Appendix A.9 for basic notations and facts related to matrices.

4.1 Matrices and graphs

Motivation. It is convenient to store graphs and study certain graph properties by using matrices and their properties. Think of a matrix as a grid where we map our graph onto. Each square in the grid corresponds to a pair of nodes in our graph. If two nodes are adjacent by an edge, then we mark that square with "1," like a little bridge between them. If they are not adjacent, that square stays blank (or 0). This grid, also called the adjacency matrix, becomes a powerful tool for understanding properties of our graph. Suppose we want to know if all your vertices can walk to each other somehow. By scanning the "bridges" in our matrix we can quickly see if there are any isolated islands of vertices stuck without connections. Similarly, the matrix can guide us if we are curious about paths between nodes. Think of tracing your finger through the "bridges" like stepping stones, following potential paths from one vertex to another. The more the bridges between them, the more the pathways!

Adjacency matrices in undirected graphs and digraphs. (1) Consider an undirected graph $G = (V, E)$. Assume that $|V| = n$. Label the vertices V as $\{1, \ldots, n\}$, i. e., identify V with $[n]$. Associate with G the matrix $A(G) = (a_{ij}) \in \{0, 1\}^{n \times n}$ as follows:

$$a_{ij} = \begin{cases} 1 & \text{if } (i, j) \in E, \\ 0 & \text{if } (i, j) \notin E. \end{cases}$$

The matrix $A(G)$ is called the *adjacency matrix* of G. Since G does not have loops, $a_{ii} = 0$, $i = 1, \ldots, n$. Note that $A(G)$ is symmetric, i. e., $a_{ij} = a_{ji}$ for all $i, j = 1, \ldots, n$ ($(A(G)^\top = A(G))$). Vice versa, any symmetric $A = (a_{ij}) \in \{0, 1\}^{n \times n}$ with zero diagonal induces a unique undirected graph $G = G(A)$ on the set of vertices $[n]$. Namely, $ij \in E$ if and only if $a_{ij} = 1$.

Example 4.1. Consider the following undirected graph G:

https://doi.org/10.1515/9783111337388-004

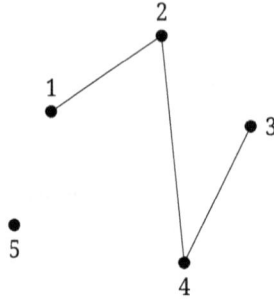

Then

$$A(G) = \begin{pmatrix} 0 & 1 & 0 & 0 & 0 \\ 1 & 0 & 0 & 1 & 0 \\ 0 & 0 & 0 & 1 & 0 \\ 0 & 1 & 1 & 0 & 0 \\ 0 & 0 & 0 & 0 & 0 \end{pmatrix}.$$

On the other hand, the symmetric matrix

$$A = \begin{pmatrix} 0 & 1 & 1 & 1 \\ 1 & 0 & 0 & 0 \\ 1 & 0 & 0 & 1 \\ 1 & 0 & 1 & 0 \end{pmatrix}$$

induces the following undirected graph:

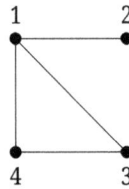

(2) Consider a digraph $G = (V, E)$. Assume that $|V| = n$. Label the vertices V as $\{1, \ldots, n\}$, i.e., identify V with $[n]$. Then with G we associate the matrix $A(G) = (a_{ij}) \in \{0, 1\}^{n \times n}$ as follows:

$$a_{ij} = \begin{cases} 1 & \text{if } (i, j) \in E, \\ 0 & \text{if } (i, j) \notin E. \end{cases}$$

The matrix $A(G)$ is called the *adjacency matrix* of the digraph G. Note that $A(G)$ may or may not be a symmetric matrix. Vice versa, any $A = (a_{ij}) \in \{0, 1\}^{n \times n}$ induces a unique digraph $G = G(A)$ on the set of vertices $[n]$. Namely, $(i, j) \in E$ if and only if $a_{ij} = 1$.

Example 4.2. Consider the following directed graph:

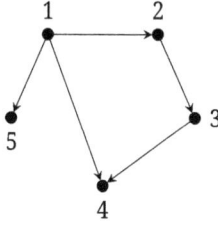

Then

$$A(G) = \begin{pmatrix} 0 & 1 & 0 & 1 & 1 \\ 0 & 0 & 1 & 0 & 0 \\ 0 & 0 & 0 & 1 & 0 \\ 0 & 0 & 0 & 0 & 0 \\ 0 & 0 & 0 & 0 & 0 \end{pmatrix}.$$

Clearly, $A(G)$ is not a symmetric matrix. Now consider the following symmetric matrix:

$$A = \begin{pmatrix} 0 & 0 & 1 & 0 & 1 \\ 0 & 0 & 1 & 0 & 1 \\ 1 & 1 & 0 & 0 & 1 \\ 0 & 0 & 0 & 0 & 1 \\ 1 & 1 & 1 & 1 & 0 \end{pmatrix}.$$

Then A induces the following directed (maximal directed) graph:

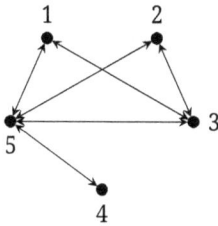

Let $A \in \{0, 1\}^{n \times n}$ be a symmetric matrix with zero diagonal. We may view $G(A)$ either as an undirected graph or as the maximal directed graph corresponding to some undirected graph. For example, consider the following symmetric matrix:

$$A = \begin{pmatrix} 0 & 0 & 1 & 0 & 1 \\ 0 & 0 & 1 & 0 & 1 \\ 1 & 1 & 0 & 0 & 1 \\ 0 & 0 & 0 & 0 & 1 \\ 1 & 1 & 1 & 1 & 0 \end{pmatrix}.$$

Then A induces the following undirected and directed graphs:

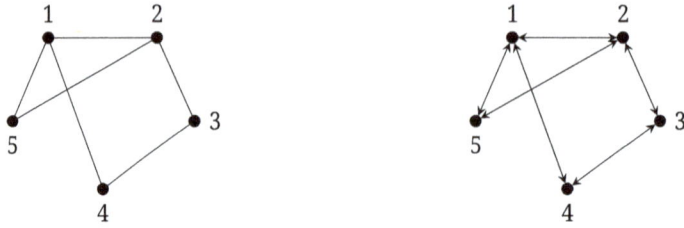

After a brief historical note on adjacency matrices, we will delve deeper into how to interpret the adjacency matrix of a graph. Additionally, we will explore the graph-theoretical properties, including finding degrees of vertices, counting walks, determining the existence of odd cycles in undirected graphs and digraphs, and examining the strong connectivity of digraphs.

Historical note. The concept of the adjacency matrix in graph theory cannot be attributed to a single individual; rather, it evolved over time as part of the development of graph theory itself. The idea of representing graphs using matrices has roots dating back to the mid-twentieth century, with numerous mathematicians contributing to its refinement. Dénes König (Figure 4.1), a Hungarian mathematician (1884–1944), played a pivotal role in this development. In the 1930s, he introduced the incidence matrix concept in his pioneering book on graph theory, "Theorie der endlichen und unendlichen Graphen" (Theory of Finite and Infinite Graphs). Although he did not explicitly use the term "adjacency matrix," König described a similar concept, utilizing a matrix to portray connections between vertices. Further advancements in the formalization of the adjacency matrix were made by Garrett Birkhoff (Figure 4.2), an American mathematician (1911–1996). In his 1946 paper titled "Tres observaciones sobre el álgebra lineal" (Three Observations on Linear Algebra), Birkhoff formally introduced the term "adjacency ma-

Figure 4.1: Dénes König (1884–1944). Photo credit: Wikipedia (In 1928). Source: https://en.wikipedia.org/wiki/D%C3%A9nes_K%C5%91nig.

Figure 4.2: Garrett Birkhoff (1911–1996). Credit photo: Wikipedia (in 1911). Source: https://en.wikipedia.org/wiki/Garrett_Birkhoff.

trix" and investigated its properties, contributing significantly to the establishment of this fundamental concept in graph theory.

We now show how to *read* some graph properties from the associated adjacency matrix.

Degrees of vertices. Let $A = (a_{ij})_{i,j=1}^{m,n} \in \mathbb{C}^{m \times n}$. Then the ith row and ith column sums of A are defined as, respectively,

$$r_i := \sum_{j=1}^{n} a_{ij}, \quad c_j := \sum_{i=1}^{m} a_{ij}, \quad j = 1, \ldots, n, \ i = 1, \ldots, m.$$

Let A be a square matrix.

- If A is symmetric, then $r_i = c_i$.
- If $A \in \{0,1\}^{n \times n}$ is symmetric with zero diagonal, then A induces an undirected graph G on n vertices, and $r_i = \deg(i)$ is the degree of the vertex i.
- If $A \in \{0,1\}^{n \times n}$ with zero diagonal, then A induces a digraph G on n vertices, and $r_i = \deg_{\text{out}}(i)$, $c_i = \deg_{\text{in}}(i)$.

Example 4.3. Let us start with the following symmetric matrix:

$$A = \begin{pmatrix} 0 & 1 & 1 & 1 & 1 \\ 1 & 0 & 1 & 0 & 0 \\ 1 & 1 & 0 & 1 & 0 \\ 1 & 0 & 1 & 0 & 1 \\ 1 & 0 & 0 & 1 & 0 \end{pmatrix}.$$

Then A induces the undirected graph $G = ([5], E)$, where

$\deg(1) = $ The sum of entries in row (or column) 1: $0 + 1 + 1 + 1 + 1 = 4$.

Similarly, $\deg(2) = 2$, $\deg(3) = 3$, $\deg(4) = 3$, and $\deg(5) = 2$.
Next, let us consider the following nonsymmetric matrix:

$$B = \begin{pmatrix} 0 & 1 & 1 & 1 & 0 \\ 0 & 0 & 0 & 1 & 1 \\ 0 & 1 & 0 & 0 & 1 \\ 1 & 0 & 0 & 0 & 0 \\ 1 & 1 & 1 & 1 & 0 \end{pmatrix}.$$

Then B induces the digraph $G = ([5], E)$, where

$\deg_{in}(1) =$ The sum of entries in column 1: $0 + 0 + 0 + 1 + 1 = 2$,

$\deg_{out}(1) =$ The sum of entries in row 1: $0 + 1 + 1 + 1 + 0 = 3$.

Similarly, $\deg_{in}(2) = 3$, $\deg_{in}(3) = 2$, $\deg_{in}(4) = 3$, $\deg_{in}(5) = 2$ and $\deg_{out}(2) = 2$, $\deg_{out}(3) = 2$, $\deg_{out}(4) = 1$, $\deg_{out}(5) = 4$.

Counting the number of walks and cycles. With the adjacency matrix of a directed or undirected graph G at hand, by fixing two vertices i and j of G, we initially determine the number of walks of length k from i to j. Subsequently, we proceed to count the number of closed walks of length k on G in accordance with the following lemma.

Lemma 4.4. *Let G be a directed or undirected graph on n vertices. Let $A = (a_{ij}) \in \{0,1\}^{n \times n}$ be the adjacency matrix of G. For $k \in \mathbb{N}$, let $A^k := (a_{ij}^{(k)})_{i,j=1}^n$. Then $A^k \in \mathbb{Z}_+^{n \times n}$, and $a_{ij}^{(k)}$ is the number of walks on G from the vertex i to the vertex j of length k. In particular, $\operatorname{tr} A^k$ is the number of closed walks on G of length k.*

Proof. Notice that $A^k = AA^{k-1}$ for each $k \in \mathbb{N}$. It follows from the definition of matrix multiplication that

$$a_{ij}^{(k)} = \sum_{l=1}^n a_{il} a_{lj}^{(k-1)}, \quad i, j = 1, \dots, n. \tag{4.1.1}$$

We first show by induction that the entries of any A^k, $k \in \mathbb{N}$, are nonnegative integers. Since $A \in \{0,1\}^{n \times n}$, each entry of A is either 0 or 1 and hence in $\mathbb{Z}_+$. Assume by induction that the result holds for $k = p$. Let $k = p+1$ and use formula (4.1.1) for this k. Since sums and products of nonnegative integers are nonnegative integers, it follows that every entry of A^{p+1} is a nonnegative integer.

We now show that the number of walks from i to j of length k in G is given by $a_{ij}^{(k)}$. Assume first that G is a digraph. For $k = 1$, this is equivalent to the definition of the adjacency matrix $A = A(G)$ of the digraph G. Assume by induction that the result holds for $k = p$. Let $k = p+1$. Let $W = \{i_0 = i, i_1, \dots, i_{p+1} = j\}$ be a walk on G of length $p+1$. Note that such a walk exists if and only if $a_{ii_1} a_{i_1 i_2} \dots a_{i_p j} = 1$ if and only if $a_{ii_1} a_{i_1 i_2} \dots a_{i_p j} \neq 0$. Denote by $\mathcal{W}_{p+1}(i,j) \supset \mathcal{W}_{p+1}(i,l,j)$ the set of all walks on G from i to j in $p+1$ steps and the subset of these walks such that the first step is from i to l. Note that either of these sets can be empty: $\mathcal{W}_{p+1}(i,j) = \emptyset$ if and only if there is no walk in G of length $p+1$ from i to j.

Then $\mathcal{W}_{p+1}(i,j) = \cup_{l=1}^{n}\mathcal{W}_{p+1}(i,l,j); \mathcal{W}_{p+1}(i,l,j) \neq \emptyset$ if and only if $a_{il} = 1$ and $\mathcal{W}_{p}(l,j) \neq \emptyset$. By the induction hypothesis $|\mathcal{W}_{p}(l,j)| = a_{lj}^{(p)}$. Hence $|\mathcal{W}_{p+1}(i,l,j)| = a_{il}a_{lj}^{(p)}$. Thus

$$|\mathcal{W}_{p+1}(i,j)| = \sum_{l=1}^{n}|\mathcal{W}_{p+1}(i,l,j)| = \sum_{l=1}^{n}a_{il}a_{lj}^{(p)} = a_{ij}^{(p+1)}.$$

Therefore $|\mathcal{W}_{k}(i,j)| = a_{ij}^{(k)}$, as claimed. In particular $|\mathcal{W}_{k}(i,i)| = a_{i}^{(k)}$ is the number of walks starting and ending at i. Hence $\operatorname{tr} A^{k}$ is the number of closed walks on G of length k.

For an undirected graph, the proof is similar, since the walk on $G = ([n],E)$ corresponds to the walk on the oriented graph where $a_{ij} = a_{ji} = 1$ if and only if $ij \in E$.   □

Example 4.5. Let $G = ([8],E)$ be an undirected graph with adjacency matrix

$$A(G) = \begin{pmatrix} 0 & 1 & 1 & 1 & 0 & 1 & 1 & 0 \\ 1 & 0 & 1 & 1 & 1 & 0 & 0 & 0 \\ 1 & 1 & 0 & 0 & 0 & 1 & 1 & 1 \\ 1 & 1 & 0 & 0 & 1 & 1 & 1 & 0 \\ 0 & 1 & 0 & 1 & 0 & 1 & 1 & 0 \\ 1 & 0 & 1 & 1 & 1 & 0 & 0 & 1 \\ 1 & 0 & 1 & 0 & 1 & 0 & 0 & 0 \\ 0 & 0 & 1 & 1 & 0 & 1 & 0 & 0 \end{pmatrix}.$$

We would like to determine all walks of length from the vertex 1 to the vertex 7. For this sake, we need to find $A(G)^{3}$:

$$A(G)^{3} = \begin{pmatrix} 10 & 14 & 13 & 16 & 7 & 17 & 5 & 7 \\ 14 & 6 & 15 & 13 & 13 & 8 & 6 & 7 \\ 14 & 15 & 9 & 10 & 9 & 16 & 12 & 11 \\ 17 & 13 & 11 & 10 & 13 & 14 & 9 & 12 \\ 7 & 13 & 8 & 12 & 4 & 15 & 13 & 6 \\ 18 & 8 & 7 & 14 & 16 & 10 & 7 & 10 \\ 12 & 4 & 11 & 8 & 11 & 5 & 3 & 6 \\ 8 & 10 & 12 & 14 & 6 & 13 & 5 & 5 \end{pmatrix}.$$

Then the number of walks of length 3 from 1 to 7 is 5. Furthermore, the number of closed walks of length 3 (3-cycles) is

$$\operatorname{tr} A(G)^{3} = 10 + 6 + 9 + 10 + 4 + 10 + 3 + 5 = 57.$$

We now determine the existence of odd cycles in graphs.

Lemma 4.6. *Let G be a directed or undirected graph on n vertices with adjacency matrix A. Then G has a cycle of odd length if and only if $\operatorname{tr} A^{2k-1}$ is at least 1 for some $k = 1, \ldots, \lfloor\frac{n+1}{2}\rfloor$.*

Proof. Assume first that G is a digraph and that there exists an odd cycle of length $2k - 1$ ($\leq n$). Clearly, $k \leq \lfloor \frac{n+1}{2} \rfloor$. Hence tr A^{2k-1} is at least 1. Suppose now that tr A^{2k-1} is at least 1 for some $k \leq \lfloor \frac{n+1}{2} \rfloor$. Hence there exists a closed walk W on G of length $2k - 1$. It is straightforward to show that any closed walk W on G can be decomposed as a "sum" of cycles. The length of the walk W is the sum of the lengths of the cycles. Since the length of the closed walk A is odd, there must be at least one odd cycle in any decomposition of W to a "sum" of cycles.

Assume now that $G = ([n], E)$ is an undirected graph. Then a *semicycle* on G is defined as a closed walk of length 2: $W = (iji)$ where $ij \in E$. Then any closed walk on G decomposes to "sum" of cycles and semicycles, and the proof in this case follows as for the digraph. □

Remark 4.7. Recall that an undirected or directed graph $G = ([n], E)$ has no odd cycles if and only if it is bipartite. (See Theorem 2.14 for the undirected graph case). It is feasible to find a fast polynomial (quadratic) algorithm in n that finds if G is bipartite or not. However, to find if a given undirected or directed graph $G = ([n], E)$ has an even cycle is a hard (NP-complete) problem.

Connectivity. We now shift our focus to examining the correlation between the adjacency matrix of a given digraph and its strong connectivity.

Lemma 4.8. *Let G be a digraph on n vertices with adjacency matrix $A \in \{0,1\}^{n \times n}$. Then G is strongly connected if and only if all the entries of $B = (b_{ij})_{i,j=1} := A^1 + \cdots + A^{n-1}$ are positive.*

Proof. Notice that $A^k \in \mathbb{Z}_+^{n \times n}$ for each $k \in \mathbb{Z}_+$. Hence $B \geq A^k$ for all $k \in [n-1]$. Assume first that G is strongly connected. Then for any two distinct vertices $i \neq j \in [n]$, there exists a path P of length k that connects i to j. Since all the vertices in P are distinct, $k \leq n - 1$. So $b_{ij} \geq a_{ij}^{(k)} \geq 1$.

Assume now that all the off-diagonal entries of B are positive. Let $i \neq j$ be two distinct vertices. From the definition of B it follows that there exists $k \in [n-1]$ such that $a_{ij}^{(k)} > 0$. Since $a_{ij}^{(k)}$ is a nonnegative integer, it follows that $a_{ij}^{(k)} \geq 1$. Thus we have a walk of length k from i to j. Hence we have a path of length at most k from i to j. Thus G is strongly connected. □

Example 4.9. Let $G = ([4], E)$ be a digraph with adjacency matrix

$$A(G) = \begin{pmatrix} 0 & 1 & 0 & 0 \\ 0 & 0 & 1 & 1 \\ 1 & 0 & 0 & 0 \\ 1 & 0 & 0 & 0 \end{pmatrix}.$$

Our goal is to determine whether G is strongly connected or not. For this sake, we employ Lemma 4.8. Setting $A := A(G)$, we have

$$B = A^1 + A^2 + A^3 = \begin{pmatrix} 2 & 1 & 1 & 1 \\ 2 & 2 & 1 & 1 \\ 1 & 1 & 1 & 1 \\ 1 & 1 & 1 & 1 \end{pmatrix}.$$

Since all entries of B are positive, we conclude that G is strongly connected. Note that $A(G)$ induces the following digraph, which is evidently strongly connected as expected:

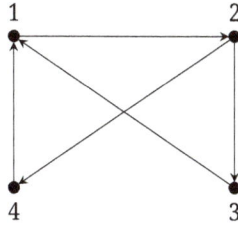

4.2 Block matrices and digraphs

In this section, we assume familiarity with certain basic concepts in linear algebra and matrices. The readers are encouraged to refer to Appendix A.9 for a quick overview of this background. See also [16] for a more detailed discussion.

$A \in \mathbb{C}^{m \times n}$ is called a *block* matrix if $A = (A_{ij})_{i,j=1}^{pq}$ and each $A_{ij} \in \mathbb{C}^{m_i \times n_j}$ for $i = 1, \ldots, p,$ $j = 1, \ldots q$:

$$A = \begin{pmatrix} A_{11} & A_{12} & \cdots & A_{1q} \\ A_{21} & A_{22} & \cdots & A_{2q} \\ \vdots & \vdots & \vdots & \vdots \\ A_{p1} & A_{p2} & \cdots & A_{pq} \end{pmatrix}, \tag{4.2.1}$$

$$A_{ij} \in \mathbb{C}^{m_i \times n_j}, \quad i = 1, \ldots, p, \quad j = 1, \ldots, q, \quad \sum_{i=1}^{p} m_i = m, \quad \sum_{j=1}^{q} n_j = n.$$

A square matrix $A \in \mathbb{C}^{n \times n}$ is called *block diagonal* if it is a block matrix where each diagonal block is a square matrix and all off-diagonal elements are zero matrices. The *direct sum of matrices* is a special type of block matrix. In particular, the direct sum of square matrices is a block diagonal matrix. We denote by $A \oplus B$ the direct sum matrices A and B. For example, if

$$A = \begin{pmatrix} a_{11} & a_{12} \\ a_{21} & a_{22} \end{pmatrix}, \quad B = \begin{pmatrix} b_{11} & b_{12} & b_{13} \\ b_{21} & b_{22} & b_{23} \\ b_{31} & b_{32} & b_{33} \end{pmatrix},$$

then

$$A \oplus B = \begin{pmatrix} A & \mathbf{0} \\ \mathbf{0} & B \end{pmatrix} = \begin{pmatrix} a_{11} & a_{12} & 0 & 0 & 0 \\ a_{21} & a_{22} & 0 & 0 & 0 \\ 0 & 0 & b_{11} & b_{12} & b_{13} \\ 0 & 0 & b_{21} & b_{22} & b_{23} \\ 0 & 0 & b_{31} & b_{32} & b_{33} \end{pmatrix}.$$

So if A is block diagonal, then

$$A = \operatorname{diag}(A_1, \ldots, A_q) = \oplus_{j=1}^{q} A_j := \begin{pmatrix} A_1 & 0 & \cdots & 0 \\ 0 & A_2 & \cdots & 0 \\ \vdots & \vdots & \vdots & \vdots \\ 0 & 0 & \cdots & A_q \end{pmatrix},$$

$$A_i \in \mathbb{C}^{n_i \times n_i}, \quad i = 1, \ldots, q, \quad \sum_{j=1}^{q} n_j = n.$$

Consider the digraph (undirected graph) G as a disjoint union of q graphs G_j, expressed as $G = \cup_{j=1}^{q} G_j$. We can now note the following observations:

- The adjacency matrix of G with respect to the adjacency matrices of G_j's is $A(G) = \oplus_{j=1}^{q} A(G_j)$.
- If $G = (V, E)$ is acyclic, then Proposition 2.27 yields that $A(G)$ is the $k \times k$ block matrix

$$A = \begin{pmatrix} 0 & A_{12} & A_{13} & A_{14} & \cdots & A_{1k} \\ 0 & 0 & A_{23} & A_{24} & \cdots & A_{2k} \\ \vdots & \vdots & \vdots & \vdots & & \vdots \\ 0 & 0 & 0 & 0 & \cdots & A_{(k-1)k} \\ 0 & 0 & 0 & 0 & \cdots & 0 \end{pmatrix},$$

$$A_{ij} \in \mathbb{C}^{n_i \times n_j}, \quad i = 1, \ldots, k, \quad j = i+1, \ldots, k. \tag{4.2.2}$$

Moreover, $A_{i(i+1)} \neq 0$ for $i = 1, \ldots, k$.

- Let the digraph $G = (V, E)$ be strongly connected and periodic. By Theorem 2.32 there is a decomposition $V = \cup_{i=1}^{p} V_i$ with $|V_i| = n_i$, $i = 1, \ldots, p$. Then the adjacency matrix $A(G)$ is given as the following $p \times p$ block matrix:

$$A = \begin{pmatrix} 0 & A_{12} & 0 & 0 & \cdots & 0 \\ 0 & 0 & A_{23} & 0 & \cdots & 0 \\ \vdots & \vdots & \vdots & \vdots & & \vdots \\ 0 & 0 & 0 & 0 & \cdots & A_{(p-1)p} \\ A_{p1} & 0 & 0 & 0 & \cdots & 0 \end{pmatrix}, \tag{4.2.3}$$

$$A_{i(i+1)} \in \mathbb{C}^{n_i \times n_{i+1}}, \quad i = 1, \ldots, p, \quad (p+1 \equiv 1). \tag{4.2.4}$$

  –   Any directed (undirected) G is bipartite if it has an adjacency matrix of the above form for $p = 2$. In other words, a bipartite graph G has the adjacency matrix

$$\begin{pmatrix} 0 & A_{12} \\ A_{21} & 0 \end{pmatrix}.$$

Reduced graphs and reducible matrices. We begin with reducible matrices. Let $A \in \mathbb{R}_+^{n \times n}$ be a nonnegative matrix. Then A induces the digraph $G(A) = ([n], E)$ with $ij \in E$ if and only if $a_{ij} > 0$. A matrix $A \in \mathbb{R}_+^{n \times n}$ is called *irreducible* if $G(A)$ is strongly connected. Otherwise, A is called *reducible*. For instance, let A be the matrix given in Example 4.9. Since $G(A)$ is strongly connected, A is irreducible. Irreducible matrices are treated with more depth and detail in Chapter 5. A *permutation* matrix is the result of repeatedly interchanging the rows and columns of an identity matrix. The set of all $n \times n$ permutation matrices is denoted by $\mathcal{P}_n$.

Theorem 4.10. *Let $A \in \mathbb{R}_+^{n \times n}$. Then there exists a permutation matrix $P \in \mathcal{P}_n$ such that $B = PAP^T$ has the following form:*

$$B = \begin{pmatrix}
B_{11} & B_{12} & \cdots & B_{1t} & B_{1(t+1)} & B_{1(t+2)} & \cdots & B_{1(t+f)} \\
0 & B_{22} & \cdots & B_{2t} & B_{2(t+1)} & B_{2(t+2)} & \cdots & B_{2(t+f)} \\
\vdots & \vdots & \ddots & \vdots & \vdots & \vdots & \ddots & \vdots \\
0 & 0 & \cdots & B_{tt} & B_{t(t+1)} & B_{t(t+2)} & \cdots & B_{t(t+f)} \\
0 & 0 & \cdots & 0 & B_{(t+1)(t+1)} & 0 & \cdots & 0 \\
\vdots & \vdots & \ddots & \vdots & \vdots & \vdots & \ddots & \vdots \\
0 & 0 & \cdots & 0 & 0 & 0 & \cdots & B_{(t+f)(t+f)}
\end{pmatrix},$$

where $B_{ij} \in \mathbb{R}^{n_i \times n_j}$, $i, j = 1, \ldots, t + f$, $n_1 + \cdots + n_{t+f} = n$, $t \geq 0$, $f \geq 1$. Each B_{ii} is irreducible, and the submatrix $B_0 := [B_{ij}]_{t+f\ i=j=t+1}$ is block diagonal. If $t = 0$, then B is block diagonal. If $t \geq 1$, then for each $i = 1, \ldots, t$, not all the matrices $B_{i(i+1)}, \ldots, B_{i(i+f)}$ are zero matrices.

Proof. Let $G_{rdc} = (V_{rdc}, E_{rdc})$ be the reduced graph of the induced digraph $G(A) = ([n], E)$. Then G_{rdc} is acyclic. Let $\ell \geq 1$ be the length of the longest path in the digraph G_{rdc}. For a given vertex $v \in V_{rdc}$, let $\ell(v)$ be the length of the longest path in G_{rdc} from v. So $\ell(v) \in [0, \ell]$. For $j \in \{0, \ldots, \ell\}$, denote by V_j the set of all vertices in V_{rdc} such that $\ell(v) = j$. Since G_{rdc} is acyclic, it follows that $V_\ell, \ldots, V_0$ is a decomposition of V_{rdc} into nonempty sets. Note that if there is a directed edge in G_{rdc} from V_i to V_j, then $i > j$. Also, we always have at least one directed edge from V_i to V_{i-1} for $i = \ell, \ldots, 1$ if $\ell > 0$. Assume that $|V_j| = m^{1+\ell-j}$ for $j = 0, \ldots, \ell$. Let $M_0 = 0$ and $M_j = \sum_{i=1}^{j} m_i$ for $j = 1, \ldots, \ell$. Then we name the vertices of V_j as $\{M_{\ell-j} + 1, \ldots, M_{\ell-j} + m^{1+\ell-j}\}$ for $j = 0, \ldots, \ell$. Let $f := |V_0| = m^{\ell+1}$ and $t := |\bigcup_{j=1}^{\ell} V_j| = \sum_{j=1}^{\ell} m^j$. Note that $f \geq 1$ and $t = 0$ if and only if $\ell = 0$. Hence the adjacency matrix $A(G_{rdc})$ is strictly upper triangular. Furthermore, the last f rows of $A(G_{rdc})$ are zero rows. Recall from Chapter 2 that each vertex in V corresponds to a maximal strongly connected component of $G(A)$. Hence B_{ii} is irreducible. Note that $B_{ij} = 0$ for $j > i$ if and only if there is no directed edge from the vertex i to the vertex j in G_{rdc}. Observe that for $i \leq t$, the vertex i represents a vertex in V_k for some $k \geq 1$. Hence, for $i \leq t$, there exists $j > i$ such that $B_{ij} = 0$. □

Remark 4.11. Let $A = \mathrm{diag}(A_1, \ldots, A_q)$ be a block diagonal matrix. Then it is straightforward to show that $A^k = \mathrm{diag}(A_1^k, \ldots, A_q^k)$ for all $k \in \mathbb{N}$. Clearly, $A^0 = \mathrm{diag}(A_1^0, \ldots, A_q^0)$. It is easy to raise to the power $k \in \mathbb{N}$ a diagonal matrix $D = \mathrm{diag}(d_1, \ldots, d_n)$: $D^k = \mathrm{diag}(d_1^k, \ldots, d_n^k)$. This basic fact can be used for computing the powers of A for "most" of square matrices, since most of the square matrices are similar to a diagonal matrix, that is, for most $A \in \mathbb{C}^{n \times n}$, there exist an invertible matrix $X \in \mathbb{C}^{n \times n}$ and a diagonal matrix $\Lambda :=$ $\mathrm{diag}(\lambda_1, \ldots, \lambda_n) \in \mathbb{C}^{n \times n}$ such that $A = X \Lambda X^{-1}$. Hence $A^k = X \Lambda^k X^{-1}$ for all $k \in \mathbb{N}$.

4.3 Worked-out problems

1. The set Π_n of all probability distributions on $[n] = \{1, 2, \ldots, n\}$ is defined as

$$\Pi_n = \left\{ a = (a_1, a_2, \ldots, a_n) \in \mathbb{R}^n \; : \; a_i \geq 0 \text{ for all } i, \text{ and } \sum_i a_i = 1 \right\}.$$

Prove that Π_n is a compact set.

Solution. To prove the compactness, we need to show that Π_n is closed and bounded in $\mathbb{R}^n$. Since $0 \leq a_i \leq 1$ for all i, Π_n is a subset of $[0,1]^n$. Since $[0,1]^n$ is bounded, Π_n is bounded. Next, we show that the complement of Π_n is open. Let $a = (a_1, a_2, \ldots, a_n)$ be a point in the complement of Π_n. Then either there exists i such that $a_i < 0$, or $\sum_i a_i \neq 1$. In the first case, the open ball $B(a, |a_i|)$ does not intersect Π_n. In the second case, there exists $\epsilon > 0$ such that $|\sum_i a_i - 1| > \epsilon$. The open ball $B(a, \epsilon/n)$ does not intersect Π_n. Therefore the complement of Π_n is open, and thus Π_n is closed. Since Π_n is both closed and bounded in $\mathbb{R}^n$, it is compact by the *Heine–Borel* theorem (refer to Appendix A.7).

4.4 Problems

1. Consider two simple undirected isomorphic graphs G_1 and G_2 with n vertices. Verify the existence of an $n \times n$ invertible matrix P such that $A(G_1) = P^{-1}A(G_2)P$.
2. Let $A(G) = (a_{ij})$ be the adjacency matrix of a connected graph $G = (V, E)$. Prove that the distance between two vertices v_i and v_j is equal to k if and only if k is the smallest integer such that $a_{ij}^{(l)} = 0$.
3. Let G be a directed connected graph. Let $\lambda_1 \geq \lambda_2 \geq \cdots \geq \lambda_n$ be eigenvalues of $A(G)$. The eigenvector corresponding to λ_1 is called the *principal eigenvector* of $A(G)$. Prove that G is bipartite if and only if $-\lambda_1$ is an eigenvalue of $A(G)$, in which case the entire

spectrum is symmetric with respect to 0. If G is bipartite, then the eigenvector of $-\lambda_1$ is derived from its principal eigenvector by changing the signs of the components in one part of the bipartition.

4. Prove that $A \in \mathbb{R}_+^{n \times n}$ is irreducible if and only if $D - A$ is irreducible for any diagonal D.

5. A nonnegative matrix $P = (p_{ij})_{i,j=1}^n$ is called a *stochastic matrix* if $\sum_{j=1}^n p_{ij} = 1$ for $i = 1, \ldots, n$. Let $P \in [0,1]^{n \times n}$ be a stochastic matrix.

   (a) Prove that $\frac{1}{m} \sum_{i=0}^{m-1} P^i$ converges to a stochastic matrix Q as $m \to \infty$.

   (b) Prove that Q is a projection (i. e., $Q^2 = Q$) that satisfies $PQ = QP = Q$.

   (c) Denote by Π_n the set of all distributions $\alpha = (\alpha_1, \ldots, \alpha_n)$ on $[n]$. For any distribution $\alpha \in \Pi_n$, prove that $\lim_{m \to \infty} \frac{1}{m} \sum_{i=0}^{m-1} \alpha P^i = \alpha Q$ is a stationary distribution of P.

   This is called the *mean ergodic theorem* for stochastic matrices.

   (**Hint:** Use Theorem 4.10.)

5 Markov chains on digraphs

Outline. This chapter introduces *Markov chains*, defining their core components and distinguishing between homogeneous and inhomogeneous processes. It explores the simulation of homogeneous Markov chains and delves into *stationary distributions* within the context of random walks on graphs. The chapter investigates the existence, uniqueness, and convergence properties of stationary distributions based on graph characteristics. It then examines the spectral properties of nonnegative matrices, particularly stochastic matrices, and their relationship to Markov chain behavior (A detailed treatment concerning nonnegative matrices can be found in [24]). The chapter continues with a discussion of reversible Markov chains. It continues with a brief treatment of the Perron–Frobenius theorem. We conclude the chapter with a discussion of the mean first passage time, the mean recurrence time, and the Kemeny constant of a Markov chain.

i 5.1 Markov chains: basic definitions and properties

Warm up. Markov chain is a concept used to describe a system that transitions between different states over time. The key idea is that the probability of moving to a new state depends only on the current state and not on the history of how the system got there. Before a formal formulation, let us start with a couple of examples. Imagine a weather system that can be either sunny or rainy on any given day. This system can be modeled by a Markov chain with two states, "Sunny" and "Rainy." We can define the probability of transitioning between these states using a transition table (matrix):

Table 5.1: Weather chart (matrix).

	Sunny (next day)	Rainy (next day)
Sunny (today)	0.8	0.2
Rainy (today)	0.4	0.6

This matrix shows that:
- If it is sunny today, then there is a 80 % chance that it will be sunny tomorrow and a 20 % chance of rain.
- If it is raining today, then there is a 40 % chance that it will be sunny tomorrow and a 60 % chance of rain.

We can use this model to predict the likelihood of future weather patterns, considering only the current weather, without needing to know the entire weather history.
Here is how Markov chain works in this example:

https://doi.org/10.1515/9783111337388-005

1. *States*: These are the different conditions or situations the system can be in. In our example, the weather could be in the "sunny" state or in the "rainy" state.
2. *Transitions*: These are the movements between states. The probability of transitioning from one state to another is what the Markov chain captures. In our example, this is illustrated in Table 5.1, which will be called a *transition matrix*.
3. *Memoryless property*: This is the key feature of Markov chains. The probability of going to a new state depends only on the current state and not on what happened before.

Example 5.1. Consider a bike share program with only three stations: A, B, and C. Assume that all bicycles must be returned to one of these stations by the end of the day. Thus every day at midnight, all bikes are stationed at one of the three locations, allowing us to examine the distribution of bikes across all stations at this specific time daily. Our goal is to model the movement of bikes from midnight of one day to midnight of the next. Over a one-day period, we observe the following:

– For the bikes borrowed from station A, 40 % are returned to station A, 50 % are returned to station B, and 10 % are returned to station C.
– For the bikes borrowed from station B, 20 % are returned to station A, 30 % are returned to station B, and 50 % are returned to station C.
– For the bikes borrowed from station C, 10 % are returned to station A, 10 % are returned to station B, and 80 % are returned to station C.

The corresponding transition matrix is

$$T = \begin{pmatrix} 0.4 & 0.5 & 0.1 \\ 0.2 & 0.3 & 0.5 \\ 0.1 & 0.1 & 0.8 \end{pmatrix}.$$

Assuming that $T = (t_{ij})$, we make the following observations:
– t_{ij} is the probability of moving from the state represented by row i to the state represented by column j in a single transition.
– t_{ij} is a conditional probability, which we can write as

$t_{ij} = \textbf{Pr}(\text{next state is the state in column } j \mid \text{current state is the state in row } i).$

– Each row adds to 1. Such a row will be called a *probability vector*.
– The transition matrix represents change over one transition period; in this example, one transition is a fixed unit of time of one day.

Markov chain. We now formulate Markov chain rigorously. We begin with a couple of definitions.

A *probability vector* or *stochastic vector* is a vector with nonnegative entries that add up to one. A nonnegative matrix $P = (p_{ij})_{i,j=1}^{n}$ is called a *stochastic matrix*, or sometimes a *transition matrix* or a *probability matrix*, if $\sum_{j=1}^{n} p_{ij} = 1$ for $i = 1, \ldots, n$, that is, each row of a stochastic matrix is a probability vector. Equivalently, P is stochastic if $P \geq 0$ and $P\mathbf{e} = \mathbf{e}$, where $\mathbf{e} := (1, \ldots, 1)^T \in \mathbb{R}^n$ is the vector whose all coordinates are equal to 1. The set of all $n \times n$ stochastic matrices is denoted by $\mathcal{S}_n$.

Let $G = ([n], E)$ be a digraph. It is convenient to view the vertices $[n]$ of G as vertices $V := \{v_1, \ldots, v_n\}$ or states $S := \{s_1, \ldots, s_n\}$. Assume that $\deg_{\text{out}}(i) \geq 1$ for all $i \in [n]$.

Imagine that a particle jumps from vertex to vertex at discrete times measured by nonnegative integers $m = 0, 1, \ldots$. For each $m \in \mathbb{Z}_+$, let X_m be the random variable that gives the position of a particle at time X_m. So

$$\mathbf{Pr}(X_m = i) = \mu_i^{(m)}, \quad i = 1, \ldots, n.$$

Here $X_m = i$ means that the particle at time m is at the vertex v_i (or s_i). Hence $\mu^{(m)} := (\mu_1^{(m)}, \ldots, \mu_n^{(m)})$ is a *row* probability vector. The sequence of random variables $X_0, X_1, \ldots$ is called a *process*, a *random process*, or a *stochastic process*. Denote by $\mathbf{Pr}(X_m | X_0, X_1, \ldots, X_{m-1})$ the probability of the distribution of X_m knowing the values of $X_0, \ldots, X_{m-1}$. More specifically,

$$\mathbf{Pr}(X_m = i_m | X_0 = i_0, X_1 = i_1, \ldots, X_{m-1} = i_{m-1})$$

denotes the probability of a particle being in location $i_m \in [n]$, provided that the particle was at location i_l at time $l = 0, \ldots, m-1$.

The process $X_0, X_1, \ldots$ is called a *Markov process* or *Markov chain* if the following property holds: For each $m \geq 1$, the conditional probability $\mathbf{Pr}(X_m | X_0, X_1, \ldots, X_{m-1})$ is equal to the conditional probability $\mathbf{Pr}(X_m | X_{m-1})$. More precisely,

$$\mathbf{Pr}(X_m = i_m | X_0 = i_0, X_1 = i_1, \ldots, X_{m-1} = i_{m-1}) = \mathbf{Pr}(X_m = i_m | X_{m-1} = i_{m-1})$$

for all $i_0, \ldots, i_m \in [n]$. The exact values of $\mathbf{Pr}(X_m = i_m | X_{m-1} = i_{m-1})$ are described as follows. Let $P_m := (p_{ij,m})_{i,j=1}^n$, $m = 1, 2, \ldots$, be a sequence of $n \times n$ stochastic matrices such that $G(P_m) := ([n], E_m)$ is a subgraph of G for each $m \in \mathbb{N}$, that is, $E_m \subset E$, $m \in \mathbb{N}$. Then

$$\mathbf{Pr}(X_m = i_m | X_{m-1} = i_{m-1}) = p_{i_{m-1}i_m,m}, \quad i_{m-1}, i_m \in [n], \quad m = 1, \ldots, \tag{5.1.1}$$

that is, if we know that at time $m - 1$, the particle is at the location i_{m-1}, then at time m the particle will only be at the locations i_m where $(i_{m-1}, i_m) \in E$. This follows from the condition that $G(P_m) \subset G$. The stochasticity of P_m is equivalent to the fact that the particle at time m jumps to some vertex in G from the vertex it was at time $m - 1$. The general form (5.1.1) is called an *inhomogeneous* Markov chain or a *time-inhomogeneous* Markov chain. We claim that

$$\mu^{(m)} = \mu^{(m-1)} P_m, \quad m = 1, \ldots. \tag{5.1.2}$$

Indeed,

$$\mathbf{Pr}(X_m = j) = \sum_{i=1}^{n} \mathbf{Pr}(X_{m-1} = i)\mathbf{Pr}(X_m = j | X_{m-1} = i) = \sum_{i=1}^{n} \mu_i^{(m-1)} p_{ij,m}.$$

In particular, we have $\mu^{(m)} = \mu^{(0)} P_1 P_2 \ldots P_m$ for all $m \in \mathbb{N}$. In other words, an inhomogeneous Markov chain is a Markov chain where the probability of transitioning from one state to another depends on both the current state and the specific time step.

Example 5.2. Consider a drunk man walking along a long straight road. Let $X(n)$ denote his position at time n relative to a fixed point on the road. The transition probabilities are given by

$$\mathbf{Pr}(X(n+1) = i + 1 \mid X(n) = i) = a_n > 0,$$
$$\mathbf{Pr}(X(n+1) = i - 1 \mid X(n) = i) = 1 - a_n > 0.$$

Here we observe that the probability at a particular time n depends on n. Thus $X(n)$ is an inhomogeneous Markov chain.

Markov chain is called *homogeneous* or *time-homogeneous* if $P_m = P$ for all $m \in \mathbb{N}$ and some fixed stochastic P. In that case, P is called the *transition* matrix of the Markov chain. In other words, time-homogeneity means that the probabilities regarding transitions between states do not change with time:

$$\mathbf{Pr}(X_{n+1} = j \mid X_n = i) = \mathbf{Pr}(X_1 = j \mid X_0 = i)$$

Usually, it is assumed that $G = G(P)$. Then

$$\mu^{(m)} = \mu^{(0)} P^m, \quad m = 1, \ldots. \tag{5.1.3}$$

Consider the "sunny–rainy" weather example at the beginning of the section. We may observe that it is a homogenous Markov chain.

Historical note. The stochastic matrix, developed alongside the Markov chain by Russian mathematician Andrey Markov (Figure 5.1), was first introduced in his 1906 publication during his tenure at St. Petersburg University [31]. Originally intended for linguistic analysis and mathematical topics such as card shuffling, both Markov chains and matrices soon found applications in various fields.

Further advancements in stochastic matrices were made by scholars like Andrey Kolmogorov (Figure 5.2), who extended their scope by introducing continuous-time Markov processes [28]. By the 1950s, stochastic matrices had been applied in fields such as econometrics and circuit theory. They have been in use since the 1970s across diverse fields. From structural science and medical diagnosis to land change modeling, these matrices have proven their versatility in formal analysis.

Figure 5.1: Andrey Markov (1856–1922). Photo credit: Wikipedia, Markov in 1886, Source: https://en. wikipedia.org/wiki/Andrey_Markov.

Figure 5.2: Andrey Kolmogorov (1903–1987). Photo credit: Wikipedia, Andrey Nikolaevich Kolmogorov 25 April 1903. Source: https://en.wikipedia.org/wiki/Andrey_Kolmogorov.

5.2 Computer simulation of homogeneous Markov chains

Let $P = (p_{ij})_{i,j}^{n} \in [0,1]^{n \times n}$ be a stochastic matrix. We aim to simulate the homogeneous Markov chain generated by P on the graph $G = G(P)$. This requires a program that generates a random variable U uniformly distributed on the interval $[0,1]$. In other words, U takes values only in $[0,1]$, and $\mathbf{Pr}(U \leq t) = t$ for all $t \in [0,1]$.

Suppose a random variable X_m has value $i \in [n]$ at time $m \in \mathbb{Z}_+$. This signifies that at time m, our particle is at state s_i. For $j = 0, 1, \ldots, n$, define $q_{ij} \in [0,1]$ as follows:

$$q_{i0} = 0, \quad q_{ij} = \sum_{l=1}^{j} p_{il}, \quad \text{for } j = 1, \ldots, n, \quad i = 1, \ldots, n. \tag{5.2.1}$$

Apply the subroutine that generates the random variable U. There exists exactly one $j \in [n]$ such that $q_{i(j-1)} \leq U < q_{ij}$. (Verify this!) Set $X_{m+1} = j$. This represents the particle jumping from state s_i at time m to state s_j at time $m + 1$. See [22] for more detail.

5.3 Stationary distributions

A *random walk* is a stochastic process describing a path consisting of a sequence of random steps in a probability space:

–   Given a digraph $G = ([n], E)$ with $\deg_{\text{out}}(i) \geq 1$ for each $i = 1. \ldots, n$ and a starting point, select a neighbor at random.
–   Move to the selected neighbor and repeat the process until a termination condition is met.
–   The random sequence of vertices selected in this way is a *random walk of the graph*.

Starting from an initial vertex v_0, randomly select an adjacent vertex v_1 and move to it. Then randomly select a neighbor v_2 of v_1 and move to v_2. Continue this process.

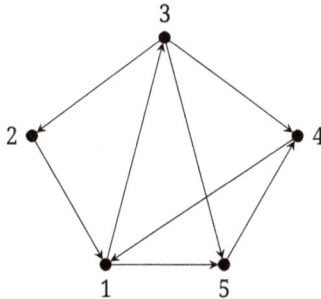

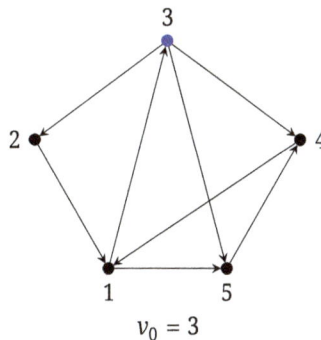

$$v_0 = 3$$

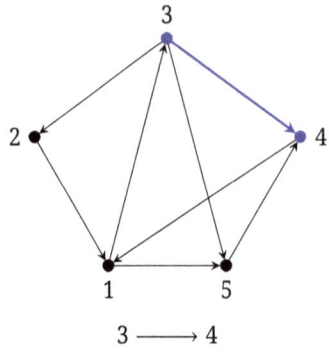

$$3 \longrightarrow 4$$

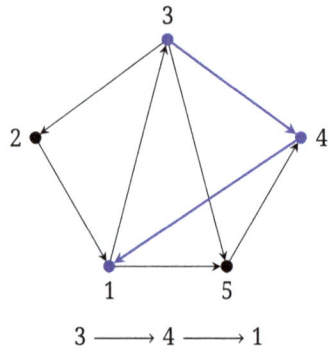

$$3 \longrightarrow 4 \longrightarrow 1$$

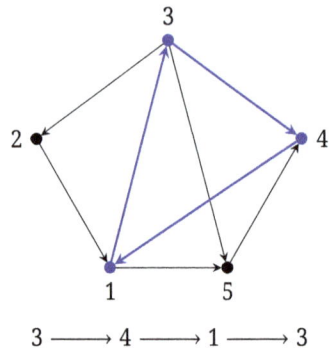

$$3 \longrightarrow 4 \longrightarrow 1 \longrightarrow 3$$

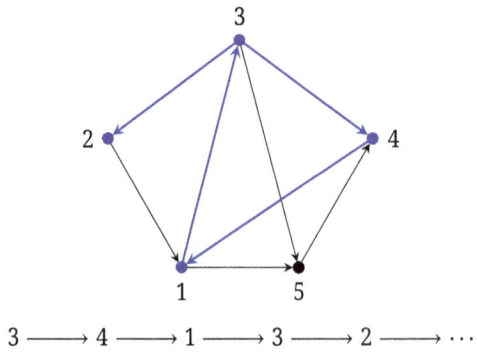

$$3 \longrightarrow 4 \longrightarrow 1 \longrightarrow 3 \longrightarrow 2 \longrightarrow \cdots$$

The sequence of vertices $v_0, v_1, v_2, \ldots, v_k, \ldots$ selected in this manner forms a simple random walk on G. At each step k, we have a random variable X_k taking values in $V(G)$. Therefore the random sequence $X_0, X_1, X_2, \ldots, X_k, \ldots$ is a stochastic process defined on the state space $V(G)$.

The probability p_{ij} that $X_{k+1} = j$ given $X_k = i$ is defined as

$$p_{ij} = \mathbf{Pr}(X_{k+1} = j \mid X_k = i) = \begin{cases} \frac{1}{\deg_{\text{out}}(i)} & \text{if } ij \in E(G), \\ 0 & \text{otherwise.} \end{cases}$$

Random walks and Markov chains. A row probability vector $\mu = (\mu_1, \ldots, \mu_n)$ is called a *stationary distribution* (or sometimes *equilibrium* or steady distribution) for a stochastic matrix $P = (p_{ij})_{i,j}^n$ if $\mu P = \mu$. This means that when the stationary distribution is multiplied by the transition matrix, it remains unchanged. In other words, μ is a stationary distribution if $\mathbf{Pr}(X_1 = i) = \mu_i$ for all $i \in [n]$ whenever $\mathbf{Pr}[X_0 = i] = \mu_i$ for all $i \in [n]$.

Consider a homogeneous Markov chain on a state space $G = G(P)$ induced by the transition matrix P. Let $X_0, X_1, \ldots$ represent the Markov process (random walk) on G defined by P. Suppose the initial distribution of X_0 is given by the stationary distribution μ: $\mathbf{Pr}(X_0 = i) = \mu_i$ for all states $i = 1, \ldots, n$. Then from equation (5.1.3) we can conclude that $\mu^{(m)} = \mu$ for any positive integer m. In other words, the random variables $X_0, X_1, \ldots$ have the same distribution.

We then raise the following natural problems in the theory of Markov chains:

1. *Do stationary distributions always exist?*
2. *Under what conditions is there a unique stationary distribution?*
3. *Suppose P admits a unique stationary distribution μ. Under what conditions does the Markov process $X_0, X_1, \ldots$ converge to a unique random variable X with the stationary distribution μ?*

The answer to the first question is yes. We will now address questions 2 and 3. To formulate the answers, we need some additional definitions, which will be introduced next.

Let $G = G(P)$. Then P is called *irreducible* if G is strongly connected. An irreducible P is called *aperiodic* (*periodic*) if G is a strongly connected aperiodic (periodic) graph. P is called *reducible* if G is not strongly connected. Consider the Markov chain consisting of three states 0, 1, and 2 with the transition matrix

$$P = \begin{pmatrix} \frac{1}{2} & \frac{1}{2} & 0 \\ \frac{1}{2} & \frac{1}{4} & \frac{1}{4} \\ 0 & \frac{1}{3} & \frac{2}{3} \end{pmatrix}.$$

It is easy to verify that this Markov chain is irreducible. For example, it is possible to go from state 0 to state 2 by the path $0 \to 1 \to 2$.

Let G_{rdc} be the reduced graph of G. Let $G(V_i)$ be a maximal strongly connected component of G. Then V_i is called a *terminal subset* if $\{V_i\}$ is a terminal vertex in G_{rdc}.

Proposition 5.3. *Let $P = (p_{ij})_{i,j=1}^{n}$ be a stochastic matrix. Let $G = G(P) = ([n], E)$ and assume that $V \subset [n]$ is a nonempty subset of $[n]$. Denote by $P(V) := (p_{ij})_{i,j \in V}$ the square submatrix of P induced by the rows and columns of P that are in V. Then $P(V)$ is a substochastic matrix, that is, each entry of $P(V)$ is nonnegative, and the sum of each row is at most 1. $P(V)$ is an irreducible stochastic matrix if and only if exactly one of the following conditions holds:*

(a) *P is irreducible, i. e., G is strongly connected. Then $V = [n]$.*

(b) *P is reducible, $G(V)$ is a maximal strongly connected component of G, and $\{V\}$ is a terminal vertex in the acyclic reduced graph G_{rdc} corresponding to G.*

Proof. Let $i \in V$. Then

$$\sum_{j \in V} p_{ij} \le \sum_{j=1}^{n} p_{ij} = 1. \tag{5.3.1}$$

Hence $P(V)$ is substochastic. Note that $\sum_{j \in V} p_{ij} = 1$ if and only if $(i, j) \notin E$ for any $j \notin V$.

Suppose first that G is strongly connected. Let $V \subsetneq [n]$ be a nonempty subset (hence $n > 1$). Since G is strongly connected, there exist $i \in V$ and $k \notin V$ such that $(i, k) \in E$. Hence $p_{ik} > 0$. In particular, (5.3.1) yields that $\sum_{j \in V} p_{ij} < 1$. Thus $P(V)$ is not stochastic. Thus we are left with the case $V = [n]$. Then $P(V) = P$ is irreducible.

Assume that P is reducible, i. e., G is not strongly connected. Suppose first that $P(V)$ is an irreducible stochastic matrix. In particular, $V \ne [n]$. Then $G(V) = G(P(V))$ is strongly connected. Since $P(V)$ is stochastic, it follows that for all $i \in V$, we have the equality sign in (5.3.1). Thus $i \in V$ and $k \notin V$, implying $(i, k) \notin V$, that is, there are no edges from V to $[n] \backslash V$. Therefore $G(V)$ is a maximal connected component of G, and V is a terminal vertex.

Suppose now that $G(V)$ is a maximal strongly connected component of G. As $G(V) = G(P(V))$, $P(V)$ is irreducible. Suppose that V is terminal. So there are no edges from V to $[n] \backslash V$. Hence for all $i \in V$, we have the equality sign in (5.3.1). Thus $P(V)$ is stochastic. □

Probabilistic interpretation of convex hull. Let $\alpha = (\alpha_1, \dots, \alpha_n)$ be a probability vector (distribution) on $[n]$. Then $\operatorname{supp} \alpha = \{i \in [n] : \alpha_i > 0\}$ is the support of the distribution α. Denote by Π_n the set of all distributions $\alpha = (\alpha_1, \dots, \alpha_n)$ on $[n]$. For any $p \in \mathbb{N}$, let $\mathbf{x}_1, \dots, \mathbf{x}_p \in \mathbb{R}^d$ be p vectors. Let $(\beta_1, \dots, \beta_p) \in \Pi_p$ be a distribution on $[p]$. Then $\mathbf{x} := \sum_{l=1}^{p} \beta_l \mathbf{x}_l \in \mathbb{R}^d$ is a convex combination of $\mathbf{x}_1, \dots, \mathbf{x}_p$. Let

$$\operatorname{conv}(\mathbf{x}_1, \dots, \mathbf{x}_p) := \left\{ \mathbf{x} \in \mathbb{R}^d : \mathbf{x} = \sum_{l=1}^{p} \beta_l \mathbf{x}_l \text{ for all } (\beta_1, \dots, \beta_p) \in \Pi_p \right\}.$$

Then conv $(\mathbf{x}_1, \ldots, \mathbf{x}_p)$ is called the *convex hull* spanned by $\mathbf{x}_1, \ldots, \mathbf{x}_p$. It is a convex set. $(C \subset \mathbb{R}^d$ is called convex set if any convex combination of any two points is in C.)

Here is a probabilistic interpretation of conv $(\mathbf{x}_1, \ldots, \mathbf{x}_p)$. Let $X : \Omega \to \mathbb{R}^d$ be a random variable such that $X(\Omega) = \{\mathbf{x}_1, \ldots, \mathbf{x}_p\}$. Then $E(X)$ is a point in $\mathbb{R}^d$. The set of all possible values of $E(X)$ is given by conv $(\mathbf{x}_1, \ldots, \mathbf{x}_p)$.

Similarly, if $a^1, \ldots, a^p \in \Pi_n$ are p distributions on $[n]$, then conv $(a^1, \ldots, a^p)$ is the convex hull of row vectors spanned by $a^1, \ldots, a^p$. It is straightforward to show that each $\alpha \in$ conv $(a^1, \ldots, a^p)$ is a distribution on $[n]$.

Theorem 5.4. *Let $P = (p_{ij})_{i,j}^n$ be a stochastic matrix. Let $G = G(P) = ([n], E)$ be the induced digraph by P. Then:*

1. *If P is irreducible, i. e., G is strongly connected, then P has a unique stationary distribution $\mu = (\mu_1, \ldots, \mu_n)$ with $\mu_i > 0$ for $i = 1, \ldots, n$.*

   (a) *If G (or P) is aperiodic, then for any distribution $\mu^{(0)}$,*

   $$\lim_{m \to \infty} \mu^{(0)} P^m = \mu.$$

   In other words, the Markov process $X_0, X_1, \ldots$, converges to the unique random variable X given by the stationary distribution μ ($\mathbf{Pr}(X = i) = \mu_i, i = 1, \ldots, n$).

   (b) *G (or P) is periodic. Assume that $p \geq 2$ is the gcd of all cycles of G. Then for each $\mu^{(0)}$, all sequences $\mu^{(0)} P^{mp+i}, m = 1, 2, \ldots$, converge for $i = 0, 1, \ldots, p - 1$. These limits depend on $\mu^{(0)}$ and i.*

2. *Assume that P is reducible, i. e., G is not strongly connected. Let G_{rdc} be the reduced graph of G with the vertices $\{V_1\}, \ldots, \{V_k\}$. Then for each terminal vertex $\{V_i\}$ in G_{rdc}, there exists a unique distribution $\mu(V_i) \in \Pi_n$ with supp $\mu(V_i) = V_i$. Assume that $\{V_{i_1}\}, \ldots, \{V_{i_t}\}$ are all the terminal vertices of G_{rdc}. Then the set of all stationary distributions of P is the convex hull spanned by $\mu(V_{i_1}), \ldots, \mu(V_{i_t})$. Hence P has a unique stationary distribution if and only if G_{rdc} has exactly one terminal vertex $\{V_{i_1}\}$.*

   (a) *Assume that G_{rdc} has exactly one terminal vertex $\{V_{i_1}\}$ and, furthermore, $G(V_{i_1})$ is aperiodic. Then for any distribution $\mu^{(0)}$,*

   $$\lim_{m \to \infty} \mu^{(0)} P^m = \mu(V_{i_1}),$$

   that is, the Markov process $X_0, X_1, \ldots$ converges to the unique random variable X given by the stationary distribution $\mu(V_{i_1})$.

   (b) *Assume that either $t = 1$ and $G(V_{i_1})$ is periodic or $t > 1$. Let $p_i \geq 1$ be the gcd of all cycles of $G(V_{ij})$ for $j = 1, \ldots, t$. Let p be the smallest positive integer divisible by $p_1, \ldots, p_t$. Then for each $\mu^{(0)}$, all sequences $\mu^{(0)} P^{mp+i}, m = 1, 2, \ldots$, converge for $i = 0, 1, \ldots, p - 1$. These limits depend on $\mu^{(0)}$ and i.*

To prove this theorem, we will need some results on the spectral properties of non-negative matrices discussed in the next section.

ℹ 5.4 Square matrices with nonnegative entries

In this section, a basic understanding of matrix-related concepts is assumed. These include various types of matrices, key parameters such as determinants and ranks, and concepts of eigenvalues, eigenvectors, and eigenbases. For a brief review of these topics, refer to Appendix A.9. See also [16] for a more detailed treatment.

We begin with eigenvalues of stochastic matrics:

Let $P \in \mathbb{R}_+^{n \times n}$, i.e., P is an $n \times n$ matrix with nonnegative entries. It is stochastic if and only if $P\mathbf{1} = \mathbf{1}$, where $\mathbf{1} = (1, \ldots, 1)^\top$.

Theorem 5.5. *Let P be a stochastic matrix. Denote by $G(P) = ([n], E)$ the directed graph induced by P. Then:*
(a) *1 is an eigenvalue of P.*
(b) *Every other eigenvalue λ of P satisfies $|\lambda| \leq 1$.*
(c) *If λ is an eigenvalue of P with modulus 1 (i. e., $|\lambda| = 1$), then there exists a terminal vertex set $\{V\}$ in the reduced graph G_{rdc} such that λ is an eigenvalue of the irreducible stochastic matrix $P(V)$ with period $p(V)$.*
(d) *λ is a simple root of $\det(zI - P(V))$, and $\lambda^{p(V)} = 1$.*
(e) *Any $p(V)$th root of 1 is an eigenvalue of $P(V)$ and hence of P.*
(f) *Any eigenvalue of P with modulus 1 is geometrically simple.*

Proof. We prove that any eigenvalue λ of P satisfies $|\lambda| \leq 1$. Suppose $P\mathbf{x} = \lambda\mathbf{x}$ for some nonzero vector $\mathbf{x} \neq \mathbf{0}$. Let $\mathbf{x} = (x_1, x_2, \ldots, x_n)^\top$ and suppose $|x_k| = \max_{i \in [1,n]} |x_i| > 0$. Then

$$|\lambda x_k| = |(P\mathbf{x})_k| = \left| \sum_{i=1}^n p_{ki} x_i \right| \leq \sum_{i=1}^n p_{ki} |x_i| \leq \sum_{i=1}^n p_{ki} |x_k| = |x_k|.$$

Dividing by $|x_k|$ gives $|\lambda| \leq 1$. The remaining claims of the theorem can be derived from the Perron–Frobenius theorem (which will be discussed in Section 5.10) together with Theorem 5.4. We will address the particular case where P is symmetric in the next section. □

5.5 Spectral theory of Hermitian matrices

Historical note. A spectral theorem over the complex numbers $\mathbb{C}$ determines when a matrix can be diagonalized, meaning that it can be represented as a diagonal matrix in some basis. This property is extremely useful because computations involving a diagonalizable matrix can often be simplified to computations involving the corresponding diagonal matrix. Note that the eigenvalues of an $n \times n$ real- or complex-valued matrix may consist of complex numbers. The spectral theorem provides a canonical decomposition, known as the spectral decomposition, of the underlying vector space on which the operator acts.

Augustin-Louis Cauchy proved the spectral theorem for symmetric matrices, demonstrating that every real symmetric matrix is diagonalizable. Additionally, Cauchy was the first to systematically study determinants. Charles Hermite extended Cauchy's results for symmetric matrices to Hermitian matrices, which are the observables in quantum physics. The spectral theorem, as generalized by John von Neumann (Figure 5.3), is now regarded as one of the most important results in operator theory. We now elaborate on the spectral theory of Hermitian matrices.

Figure 5.3: John von Neumann (1903–1957). Photo credit: Wikipedia, von Neumann in the 1940s. Source: https://en.wikipedia.org/wiki/John_von_Neumann.

Let $z = x + y\mathbf{i}$ be a complex number. Then the conjugate of z is defined as $\bar{z} := x - y\mathbf{i}$. For $A \in \mathbb{C}^{n \times n}$, the conjugate transpose is defined as $A^* := \bar{A}^\top$. A matrix $A \in \mathbb{C}^{n \times n}$ is called *Hermitian* if $A^* = A$. Let $\mathbb{H}_n$ denote the set of all $n \times n$ Hermitian matrices. Note that $\mathbb{H}_n$ forms a vector space over $\mathbb{R}$. Observe that $\mathbb{S}_n(\mathbb{R}) := \mathbb{H}_n \cap \mathbb{R}^{n \times n}$ is the set of real symmetric matrices in $\mathbb{R}^{n \times n}$. A Hermitian matrix A has only real eigenvalues and is diagonalizable.

Let $A \in \mathbb{H}_n$. It is important to note that any conjugate of the right eigenvector of A is also a left eigenvector corresponding to the same eigenvalue. Additionally, it is possible to choose the eigenvectors $\mathbf{x}_1, \ldots, \mathbf{x}_n \in \mathbb{C}^n$ of A such that

$$A\mathbf{x}_i = \lambda_i \mathbf{x}_i, \quad i = 1, \ldots, n, \quad \lambda_1 \geq \lambda_2 \geq \cdots \geq \lambda_n, \quad \mathbf{x}_i^* \mathbf{x}_j = \delta_{ij}, \quad i, j = 1, \ldots, n. \quad (5.5.1)$$

Here δ_{ij} represents the (i, j) entry of the identity matrix I_n. The set of n vectors $\mathbf{x}_1, \ldots, \mathbf{x}_n \in \mathbb{C}^n$ satisfying the conditions $\mathbf{x}_i^\top \mathbf{x}_j = \delta_{ij}$ for $i, j = 1, \ldots, n$ is called an *orthonormal basis* of $\mathbb{C}^n$. Note that if $A \in \mathbb{S}_n(\mathbb{R})$, then we can assume that $\mathbf{x}_1, \ldots, \mathbf{x}_n$ is an orthonormal basis in $\mathbb{R}^n$.

Let $X = (\mathbf{x}_1, \ldots, \mathbf{x}_n) \in \mathbb{C}^{n \times n}$. The condition that $\mathbf{x}_1, \ldots, \mathbf{x}_n$ form an orthonormal set is equivalent to $X^* X = I_n$, which holds if and only if $X^{-1} = X^*$, i. e., X is a *unitary matrix*. Denote by $\mathbb{U}_n$ the set (group) of unitary matrices. Then equation (5.5.1) is equivalent to $A = X \Lambda X^*$. Note that $X \in \mathbb{C}^{n \times n}$ is a unitary matrix if and only if $(X\mathbf{x})^* (X\mathbf{x}) = \mathbf{x}^* \mathbf{x}$ for all $\mathbf{x} \in \mathbb{C}^n$. Denote by $\mathbb{O}_n := \mathbb{U}_n \cap \mathbb{R}^{n \times n}$ the group of real orthogonal matrices. Observe that $\|\mathbf{x}\| := \sqrt{\mathbf{x}^* \mathbf{x}}$ is a nonnegative number. In particular, it is positive unless $\mathbf{x} = 0$; $\|\mathbf{x}\|$ is called the *norm* or *length* of $\mathbf{x}$.

The maximal eigenvalue λ_1 has the following characterization:

$$\lambda_1 = \max_{\mathbf{x} \neq 0} \frac{\mathbf{x}^* A \mathbf{x}}{\mathbf{x}^* \mathbf{x}} = \max_{\|\mathbf{x}\|=1} \mathbf{x}^* A \mathbf{x}. \tag{5.5.2}$$

Equality is achieved if and only if $\mathbf{x}$ is an eigenvector corresponding to λ_1. Here is a quick argument: If $A = \Lambda = \mathrm{diag}(\lambda_1, \ldots, \lambda_n)$, then the proof is straightforward. In the general case, let $\mathbf{y} = X^* \mathbf{x}$ and note that $\mathbf{x}^* A \mathbf{x} = \mathbf{y}^* \Lambda \mathbf{y}$ and $\mathbf{x}^* \mathbf{x} = \mathbf{y}^* \mathbf{y}$. The proof for this case follows from the previous case.

Similarly,

$$\lambda_n = \min_{\mathbf{x} \neq 0} \frac{\mathbf{x}^* A \mathbf{x}}{\mathbf{x}^* \mathbf{x}} = \min_{\|\mathbf{x}\|=1} \mathbf{x}^* A \mathbf{x}.$$

Equality holds if and only if $\mathbf{x}$ is the eigenvector corresponding to λ_n. There is also an extremal characterization of λ_2:

$$\lambda_2 = \max_{\mathbf{x} \neq 0, \mathbf{x}_1^* \mathbf{x} = 0} \frac{\mathbf{x}^* A \mathbf{x}}{\mathbf{x}^* \mathbf{x}} = \max_{\|\mathbf{x}\|=1, \mathbf{x}_1^\top \mathbf{x} = 0} \mathbf{x}^* A \mathbf{x}. \tag{5.5.3}$$

Equality is achieved if and only if $\mathbf{x}$ is an eigenvector corresponding to λ_2 that is orthogonal to $\mathbf{x}_1$. The disadvantage of this characterization is that it requires knowledge of $\mathbf{x}_1$. There are other characterizations that avoid this, but we will not present them here.

Let $G(A) = (V, E)$ be the graph induced by $A = \mathrm{diag}(\lambda_1, \ldots, \lambda_n)$. Let $V = \bigcup_{i=1}^k V_i$ be a decomposition of V into a union of nonempty disjoint sets such that each $G(V_i)$ is a connected component of G. Then there exists a permutation matrix $Q \in \{0, 1\}^{n \times n}$ such that

$$Q A Q^T = \mathrm{diag}(A_1, \ldots, A_k), \quad A_i \in \mathbb{C}^{n_i \times n_i}, \quad i = 1, \ldots, k. \tag{5.5.4}$$

Furthermore, each $A_i \in \mathbb{S}_{n_i}(\mathbb{R})$ is real symmetric. Assume, as in the previous section, that $Q = I_n$. Then we have the decomposition

$$A = \mathrm{diag}(A_1, \ldots, A_k) = \oplus_{i=1}^k A_i, \quad A_i \in \mathbb{C}^{n_i \times n_i}, \tag{5.5.5}$$

where $A_i \in \mathbb{S}_{n_i}(\mathbb{R})$ for $i = 1, \ldots, k$. To find each eigenspace $\mathrm{Eig}(\lambda, A)$, we use

$$\mathrm{Eig}\left(\lambda, \bigoplus_{i=1}^k A_i\right) = \bigoplus_{i=1}^k \mathrm{Eig}(\lambda, A_i) \quad \text{and} \quad \dim \mathrm{Eig}\left(\lambda, \bigoplus_{i=1}^k A_i\right) = \sum_{i=1}^k \dim \mathrm{Eig}(\lambda, A_i) \tag{5.5.6}$$

for all $\lambda \in \mathrm{spec}(A)$. (In fact, the formula holds for all $\lambda \in \mathbb{C}$.)

Recall the Gram–Schmidt process, abbreviated as GSP, in $\mathbb{C}^n$: For a given n linearly independent vectors $\mathbf{y}_1, \ldots, \mathbf{y}_n$ in $\mathbb{C}^n$, the GSP constructs efficiently an orthonormal basis

$\mathbf{x}_1, \ldots, \mathbf{x}_n$ in $\mathbb{C}^n$ such that $\mathbf{x}_1 = \frac{1}{\|\mathbf{y}_1\|}\mathbf{y}_1$; see Appendix A.9.7. We use the GSP to derive Schur's semidiagonalization theorem of Issai Schur:

Theorem 5.6. *Let $A \in \mathbb{C}^{n \times n}$ with eigenvalues $\lambda_1, \ldots, \lambda_n$, where $\det(\lambda I_n - A) = \prod_{i=1}^{n}(\lambda - \lambda_n)$. Then there exist an $n \times n$ unitary matrix U and an upper diagonal matrix C with diagonal entries $\lambda_1, \ldots, \lambda_n$ such that $A = UCU^*$.*

Proof. We prove this theorem by induction on n. For $n = 1$, the theorem is trivial. Assume that the theorem holds for $n = m \geq 1$ and assume that $n = m + 1$. As λ_1 is an eigenvalue of A, it has a nonzero eigenvector $\mathbf{x}_1 \in \mathbb{C}^n$. We can assume that $\|\mathbf{x}_1\| = 1$. Complete $\mathbf{x}_1$ to an orthogonal basis $\mathbf{x}_1, \ldots, \mathbf{x}_n$. Let $X = (\mathbf{x}_1, \ldots, \mathbf{x}_n) \in \mathbb{U}_n$, and set $C_1 = X^* A X$. Observe that C_1 has the same eigenvalues as A and $C_1 \mathbf{e} = \lambda_1 \mathbf{e}$, where $\mathbf{e} = (1, 0, \ldots, 0)^\top$. Hence C_1 has the block upper triangular form $\left(\begin{smallmatrix} \lambda_1 & \mathbf{c}^\top \\ 0 & A_1 \end{smallmatrix}\right)$. Recall that the eigenvalues of A_1 are $\lambda_2, \ldots, \lambda_n$. Apply the induction hypothesis to A_1 to deduce that there exists a unitary matrix $X_1 \in \mathbb{U}_{n-1}$ such that $X_1^* A_1 X$ is an upper triangular matrix. Let $U = X \operatorname{diag}(1, X_1)$ to deduce the theorem. □

5.6 Singular value decomposition and l_p operator norms of matrices

Historical note. Among various useful matrix decompositions, the singular value decomposition (SVD) stands out. This decomposition factors a matrix A into the product $U \Sigma V^*$, where U and V are unitary matrices, and Σ is a real diagonal matrix. The SVD is particularly significant for several reasons. Firstly, the use of unitary matrices in the decomposition makes it an excellent tool for exploring the geometry of n-dimensional space. Secondly, the SVD is stable: small perturbations in A lead to small perturbations in Σ, and vice versa. Thirdly, the diagonal nature of Σ simplifies the identification of when A is close to a rank-deficient matrix. When this is the case, the SVD provides optimal low-rank approximations of A. Finally, thanks to the pioneering work of Gene Golub (Figure 5.4), there are efficient and stable algorithms available for computing the SVD. Eugenio Beltrami (1835–1899), Camille Jordan (1838–1921), James Joseph Sylvester (1814–1897), Er-

Figure 5.4: Gene Golub (1932–2007). Photo credit: Wikipedia, Gene Golub in 2007. Source: https://en.wikipedia.org/wiki/Gene_H._Golub.

hard Schmidt (1876–1959), and Hermann Weyl (1885–1955) were instrumental in establishing the existence of the SVD and in developing its theoretical foundations.

In what follows, we state and prove one of many versions of SVD.

Theorem 5.7. *Let $A \in \mathbb{C}^{m \times n}$. Then there exist unitary matrices $U \in \mathbb{C}^{m \times m}$ and $V \in \mathbb{C}^{n \times n}$ and a real diagonal matrix $\Sigma := \mathrm{diag}(\sigma_1, \ldots, \sigma_{\min(m,n)}) \in \mathbb{R}^{m \times n}$, where $\sigma_1 \geq \sigma_2 \geq \cdots \geq \sigma_{\min(m,n)} \geq 0$, such that*

$$A = U \Sigma V^*.$$

Assume that $\sigma_1 \geq \cdots \geq \sigma_r > \sigma_{r+1} = \cdots = \sigma_{\min(m,n)} = 0$ and $\sigma_i = 0$ for $i > \min(m, n)$. Then $r = \mathrm{rank}(A)$. The values $\sigma_1^2, \ldots, \sigma_r^2$ are the positive eigenvalues of the Hermitian matrices AA^* and A^*A, listed in decreasing order. All other eigenvalues of AA^* and A^*A are zero.

Let $U = (\mathbf{u}_1, \ldots, \mathbf{u}_m)$ and $V = (\mathbf{v}_1, \ldots, \mathbf{v}_n)$. The vectors $\mathbf{u}_1, \ldots, \mathbf{u}_m$ are the orthonormal eigenvectors of AA^* corresponding to the eigenvalues $\sigma_1^2, \ldots, \sigma_m^2$. Furthermore, $\mathbf{v}_i = \frac{1}{\sigma_i} A^* \mathbf{u}_i$, $i = 1, \ldots, r$, form an orthonormal system in $\mathbb{C}^n$. The vectors $\mathbf{v}_{r+1}, \ldots, \mathbf{v}_n$ can be any vectors that complete $\mathbf{v}_1, \ldots, \mathbf{v}_r$ to an orthonormal basis $\mathbf{v}_1, \ldots, \mathbf{v}_n$ of $\mathbb{C}^n$.

Similarly, if $\mathbf{v}_1, \ldots, \mathbf{v}_n$ is any orthonormal set of eigenvectors of A^*A corresponding to the eigenvalues $\sigma_1^2, \ldots, \sigma_n^2$, then $\mathbf{u}_i = \frac{1}{\sigma_i} A \mathbf{v}_i$ for $i = 1, \ldots, r$. The vectors $\mathbf{u}_{r+1}, \ldots, \mathbf{u}_m$ can be any vectors that complete $\mathbf{u}_1, \ldots, \mathbf{u}_r$ to an orthonormal basis $\mathbf{u}_1, \ldots, \mathbf{u}_m$ of $\mathbb{C}^m$.

Proof. Consider the matrix $B = AA^* \in \mathbb{C}^{m \times m}$. Since B is Hermitian and nonnegative definite, we have

$$\mathbf{x}^* B \mathbf{x} = (A^* \mathbf{x})^* (A^* \mathbf{x}) \geq 0$$

for all vectors $\mathbf{x} \in \mathbb{C}^m$. Thus all eigenvalues of B are nonnegative. Denote these eigenvalues by $\sigma_1^2 \geq \sigma_2^2 \geq \cdots \geq \sigma_m^2 \geq 0$.

Note that if $A^* \mathbf{x} = 0$, then $B\mathbf{x} = 0$. Conversely, if $B\mathbf{x} = 0$, then

$$\mathbf{x}^* B \mathbf{x} = 0 = (A^* \mathbf{x})^* (A^* \mathbf{x}) \implies A^* \mathbf{x} = 0.$$

Thus rank $(A) = $ rank $(B) = r$. Consequently, we have $\sigma_1 \geq \cdots \geq \sigma_r > \sigma_{r+1} = \cdots = \sigma_{\min(m,n)} = 0$.

Let $\mathbf{u}_i$ be the orthonormal eigenvectors of B corresponding to the eigenvalues σ_i^2 for $i = 1, \ldots, m$. Thus

$$B\mathbf{u}_i = \sigma_i^2 \mathbf{u}_i.$$

Since

$$\sigma_i^2 = \mathbf{u}_i^T B \mathbf{u}_i = (A^* \mathbf{u}_i)^* (A^* \mathbf{u}_i),$$

it follows that

$$\mathbf{v}_i = \frac{1}{\sigma_i} A^* \mathbf{u}_i$$

for $i = 1, \ldots, r$ form an orthonormal system in $\mathbb{C}^n$. Let $\mathbf{v}_{r+1}, \ldots, \mathbf{v}_n$ be any vectors that complete $\mathbf{v}_1, \ldots, \mathbf{v}_r$ to an orthonormal basis $\mathbf{v}_1, \ldots, \mathbf{v}_n$ of $\mathbb{C}^n$.

It can be shown straightforwardly that $A = U\Sigma V^*$, where $U = (\mathbf{u}_1, \ldots, \mathbf{u}_m)$, $\Sigma = \mathrm{diag}(\sigma_1, \ldots, \sigma_r)$, and $V = (\mathbf{v}_1, \ldots, \mathbf{v}_n)$. The remaining part of the theorem follows similarly. □

Corollary 5.8. *Let $A \in \mathbb{H}_n$ and suppose that $A = X \mathrm{diag}(\lambda_1, \ldots, \lambda_n)X^*$, where $X \in \mathbb{U}_n$ is the spectral decomposition of A. Then the sequence of singular values $\sigma_1 \geq \cdots \geq \sigma_n \geq 0$ of A is the rearranged sequence of $|\lambda_1|, \ldots, |\lambda_n|$. Specifically, if $\sigma_i = |\lambda_j|$, then we can choose $\mathbf{v}_i = \mathbf{x}_j$ and $\mathbf{u}_i = \epsilon_i \mathbf{x}_j$, where $\epsilon_i = \frac{\lambda_j}{|\lambda_j|}$ if $\lambda_j \neq 0$ and $\epsilon_i = \pm 1$ if $\lambda_j = 0$.*

Proof. Using the identity $AA^* = A^*A = A^2 = X \mathrm{diag}(\lambda_1^2, \ldots, \lambda_n^2)X^*$, the singular values of A are indeed the square roots of the eigenvalues of A^*A or AA^*. These are given by $|\lambda_1|, \ldots, |\lambda_n|$ arranged in descending order. Hence the corollary follows. □

In what follows, we need the following lemma. The reader is referred to Appendix A.9.4 for definitions concerning norms.

Lemma 5.9. *Let $C \in \mathbb{C}^{n \times n}$ be an upper triangular matrix of the form $\Lambda + B$, where Λ is a diagonal matrix, and B is a strictly upper triangular matrix. Then for $p \in [1, \infty]$ and $k \in \mathbb{N}$, we have the inequality*

$$\|C^k\|_p \leq \sum_{m=0}^{\min(k,n-1)} \binom{k}{m} \rho(\Lambda)^{k-m} \|B\|_p^m. \tag{5.6.1}$$

Proof. A matrix $B = [b_{ij}] \in \mathbb{C}^{n \times n}$ is called an $\ell \in \{0, 1, \ldots, n\}$ upper triangular if $b_{ij} = 0$ for $j \leq i + \ell - 1$. Thus 0-upper triangular is upper triangular, 1-upper triangular is strictly upper triangular, and n-upper triangular is 0. Observe that if B_1 is strictly upper triangular and B_2 is ℓ-upper triangular, then $B_1 B_2$ and $B_2 B_1$ are $(\ell + 1)$-upper triangular. Hence, if $B_1, \ldots, B_m$ are strictly upper triangular and $m \geq n$, then $B_1 \cdots B_m = 0$.

Consider $(\Lambda + B)^k$, where Λ a diagonal matrix, and B is strictly upper triangular. It is a sum of 2^k products of the form $X_1 \ldots X_k$, where $X_i \in \{\Lambda, B\}$. Let $\mathcal{W}_{m,k}$ be all products of the form $X_1 \ldots X_k$, where exactly m matrices out of $X_1 \ldots X_k$ are equal to B. Note that the cardinality of $\mathcal{W}_{m,k}$ is $\binom{k}{m}$. Observe that $\Lambda^q B \Lambda^r$ is strictly upper triangular for nonnegative integers q, r. Hence, for $k \geq m \geq n$, we obtain that $X_1 \cdots X_k = 0$ if $X_1 \cdots X_k \in \mathcal{W}_{m,k}$. Thus

$$(\Lambda + B)^k = \sum_{m=0}^{\min(k,n-1)} \sum_{X_1 \cdots X_k \in \mathcal{W}_{m,k}} X_1 \cdots X_k.$$

Next, observe that if $X_1 \ldots X_m \in \mathcal{W}_{m,k}$, then

$$\|X_1 \cdots X_k\|_p \le \|X_1\|_p \cdots \|X_k\|_p = \|\Lambda\|_p^{k-m} \|B\|_p^m = \rho(\Lambda)^{k-m} \|B\|_p^m.$$

Combine inequalities (3b) in Appendix A.9.4 with the above equality and inequality to deduce (5.6.1). □

Theorem 5.10. *For any $m \times n$ complex-valued matrix $A = (a_{ij})_{i,j=1}^{m,n}$, we have*

$$\|A\| := \|A\|_2 = \sigma_1(A), \quad \|A\|_1 = \max_{j \in [1,n]} \sum_{i=1}^{m} |a_{ij}|, \quad \|A\|_\infty = \max_{i \in [1,m]} \sum_{j=1}^{n} |a_{ij}|.$$

Proof. Clearly, $\|Q\mathbf{x}\| = \|\mathbf{x}\|$ for every unitary matrix Q. Use singular value decomposition of A to deduce

$$\|A\| = \|U \operatorname{diag}(\sigma_1, \ldots, \sigma_{\min(m,n)}) V^*\|$$
$$= \|\operatorname{diag}(\sigma_1, \ldots, \sigma_{\min(m,n)}) V^*\| = \|\operatorname{diag}(\sigma_1, \ldots, \sigma_{\min(m,n)})\| = \sigma_1.$$

For $\mathbf{x} = (x_1, \ldots, x_n)^\top \in \mathbb{C}^n$, we have

$$\|A\mathbf{x}\|_1 = \sum_{i=1}^{m} \left| \sum_{j=1}^{n} a_{ij} x_j \right| \le \sum_{i,j=1}^{m,n} |a_{ij}| \, |x_j| = \sum_{j=1}^{n} \left(\sum_{i=1}^{m} |a_{ij}| \right) |x_j| \le \left(\max_{j \in [1,n]} \sum_{i=1}^{m} |a_{ij}| \right) \|\mathbf{x}\|_1.$$

Hence $\|A\|_1 \le \max_{j \in [1,n]} \sum_{i=1}^{m} |a_{ij}|$. Choose $\mathbf{x} = \mathbf{e}_k = (\delta_{1k}, \ldots, \delta_{nk})^\top$, where $k = \arg\max_{j \in [1,n]} \sum_{i=1}^{m} |a_{ij}|$, i.e., an index for which the maximum is achieved. Then $\|A\mathbf{e}_k\|_1 = \max_{j \in [1,n]} \sum_{i=1}^{m} |a_{ij}| \, \|\mathbf{e}_k\|_1 = \max_{j \in [1,n]} \sum_{i=1}^{m} |a_{ij}|$. Hence the second equality of the theorem follows. The third equality of the theorem follows similarly. □

Corollary 5.11. *Let $A \in \mathbb{C}^{m \times n}$. Then $\|A^*\|_1 = \|A\|_\infty$. Let $P \in [0,1]^{n \times n}$ be a stochastic matrix. Then $\|P\|_\infty = \|P^\top\|_1 = 1$. In particular, $\|P\mathbf{x}\|_\infty \le \|\mathbf{x}\|_\infty$ and $\|\mathbf{y}^\top P\|_1 \le \|\mathbf{y}^\top\|_1$ for all $\mathbf{x}, \mathbf{y} \in \mathbb{R}^n$. Equality holds for all $\mathbf{y} \in \mathbb{R}_+^n$.*

Definition 5.12. *A sequence of real numbers $a_i, i \in \mathbb{N}$ is called subadditive if $a_{i+j} \le a_i + a_j$ for all $i,j \in \mathbb{N}$.*

The following lemma is fundamental in many areas of mathematics and is called *Fekete's subadditive lemma*, named after Michael Fekete (1886–1957), a Hungarian mathematician and set theorist, who worked on the transfinite diameter of a set.

Lemma 5.13. *Let $a_i, i \in \mathbb{N}$, be a real subadditive sequence. Assume that $\frac{a_i}{i}, i \in \mathbb{N}$, is bounded from below. Then the sequence $\frac{a_i}{i}$ converges to a limit $a \in \mathbb{R}$, and $a \le \frac{a_j}{j}$ for each $j \in \mathbb{N}$.*

Proof. Assume that $a_0 = 0$. Then the subadditivity conditions extend to $a_i, i \in \mathbb{Z}_+$. Note that $a_{ij} = a_{j+(i-1)j} \le a_j + a_{(i-1)j} \le \cdots \le i a_j$. Hence $\frac{a_{ij}}{ij} \le \frac{a_j}{j}$. Let $a = \liminf_{i \to \infty} \frac{a_i}{i}$. Since $\frac{a_i}{i}, i \in \mathbb{N}$, is bounded from below, $a \in \mathbb{R}$. Assume that $n_j \in \mathbb{N}, j \in \mathbb{N}$, are increasing sequences of integers such that $\lim_{j \to \infty} \frac{a_{n_j}}{n_j} = a$. Fix $\epsilon > 0$ and assume that $\frac{a_{n_j}}{n_j} \le a + \epsilon$

for some i big enough. Let $k \geq n_j$. Then $k = in_j + r$ for some $i \geq 1$ and $r \in [0, n_j - 1]$. Use subadditivity to conclude that $a_k \leq a_{in_j} + a_r \leq ia_{n_j} + a_r \leq in_j(a + \epsilon) + a_r$. This inequality yields $\lim \sup_{k \to \infty} \frac{a_k}{k} \leq a + \epsilon$. Since ϵ is an arbitrary small positive number, it follows that

$$a = \lim_{i \to \infty} \inf \frac{a_i}{i} \leq \lim_{k \to \infty} \sup \frac{a_k}{k} \leq a \Rightarrow \lim_{i \to \infty} \frac{a_i}{i} = a.$$

Since $\frac{a_{ij}}{ij} \leq \frac{a_j}{j}$, it follows that $a = \lim_{i \to \infty} \frac{a_{ij}}{ij} \leq \frac{a_j}{j}$. □

Remark 5.14. The lemma also holds in the case that $\frac{a_i}{i}, i \in \mathbb{N}$, is not bounded from below. Then $a = -\infty$. ❗

Theorem 5.15. *Let $A \in \mathbb{C}^{n \times n}$, and let $\rho(A)$ denote the spectral radius of A. Then for all $k \in \mathbb{N}$ and $p \in [1, \infty]$, we have*

$$\rho(A) \leq \|A^k\|_p^{\frac{1}{k}}.$$

Furthermore, we have

$$\lim_{k \to \infty} \|A^k\|_p^{\frac{1}{k}} = \rho(A)$$

for all $p \in [1, \infty]$ and $A \in \mathbb{C}^{n \times n}$.

Proof. Let λ be an eigenvalue of A. Then $Ax = \lambda x$ for some $x \neq 0$. Consequently,

$$\lambda^k x = A^k x \Rightarrow \|\lambda^k x\|_p = |\lambda|^k \|x\|_p = \|A^k x\|_p \leq \|A^k\|_p \|x\|_p \Rightarrow |\lambda|^k \leq \|A^k\|_p \Rightarrow \rho(A) \leq \|A^k\|_p^{\frac{1}{k}}.$$

To show the equality $\lim_{k \to \infty} \|A^k\|_p^{\frac{1}{k}} = \rho(A)$, we use that $A = UCU^* = UCU^{-1}$ by Theorem 5.6. Here $C = \Lambda + B$, where $\Lambda = \mathrm{diag}(\lambda_1, \ldots, \lambda_n)$, B is a strictly upper triangular matrix, and $\lambda_1, \ldots, \lambda_n$ are the eigenvalues of A. Clearly, $\rho(A) = \rho(C)$, and $A^k = UC^k U^*$. Hence

$$\rho(A)^k \leq \|A^k\|_p \leq \|U\|_p \|C^k\|_p \|U^{-1}\|_p$$
$$\Rightarrow \rho(A) \leq \|A^k\|_p^{1/k} \leq \|U\|_p^{1/k} \|C^k\|_p^{1/k} \|U^{-1}\|_p^{1/k}$$
$$\Rightarrow \rho(A) \leq \lim_{k \to \infty} \sup \|A^k\|_p^{1/k} \leq \lim_{k \to \infty} \sup \|C^k\|_p^{1/k}.$$

Thus it suffices to prove the theorem for $C = \Lambda + B$. We now use inequality (5.6.1) for $k > n - 1$. Then

$$\rho(A) \leq \|C^k\|_p^{1/k} \leq \rho(A)^{(k-n+1)/k} \left(\sum_{m=0}^{n-1} \binom{k}{m} \rho(\Lambda)^{n-1-m} \|B\|_p^m \right)^{1/k}.$$

Let $k \to \infty$ to deduce the theorem. □

5.7 Nonnegative symmetric matrices

This section focuses on nonnegative symmetric matrices and examines the relationship between the graph-theoretic properties of their associated graphs and their eigenvalues. A more detailed discussion can be found in [5]. Denote by $\mathbb{S}_n(\mathbb{R}_+)$ the set of all $n \times n$ real symmetric matrices with nonnegative entries. In this section, we assume that $A \in \mathbb{S}_n(\mathbb{R}_+)$.

Theorem 5.16. *Let $A = (a_{ij})_{i,j=1}^n \neq 0$ be a real symmetric matrix with nonnegative entries, and assume that the graph $G(A)$ is connected. Arrange the eigenvalues of A in decreasing order as $\lambda_1 \geq \lambda_2 \geq \cdots \geq \lambda_n$.*
Then:
1. *$\lambda_1 > 0$ and $\lambda_1 \geq |\lambda_n|$, i.e., $\lambda_1 = \rho(A)$, where $\rho(A)$ is the spectral radius of A.*
2. *$\lambda_1 > \lambda_2$, which means that λ_1 is a simple root of $\det(zI_n - A)$.*
3. *The corresponding eigenvector $\mathbf{x}_1$ in (5.5.1) can be chosen to be a unit vector with positive coordinates.*
4. *$\lambda_1 > |\lambda_n|$ unless $G(A)$ is bipartite. If $G(A)$ is bipartite, then $\lambda_n = -\lambda_1$, and $\lambda_1 > |\lambda_i|$ for $i = 2, \ldots, n-1$.*

Proof. For any vector $\mathbf{x} = (x_1, \ldots, x_n)^\top \in \mathbb{R}^n$, let $|\mathbf{x}| = (|x_1|, \ldots, |x_n|)^\top$ be the vector with nonnegative coordinates. Note that $\mathbf{x}^\top \mathbf{x} = |\mathbf{x}|^\top |\mathbf{x}|$.

Since all entries of A are nonnegative, for all $\mathbf{x} \in \mathbb{R}^n$, we have

$$|\mathbf{x}^\top A \mathbf{x}| = \left| \sum_{i,j=1}^n a_{ij} x_i x_j \right| \leq \sum_{i,j=1}^n a_{ij} |x_i| |x_j| = |\mathbf{x}|^\top A |\mathbf{x}|.$$

Thus in the maximum characterization of (5.5.2), it is sufficient to consider vectors $\mathbf{x}$ with nonnegative coordinates.

Let $\mathbf{e} = (1, \ldots, 1)^\top$. Then

$$\lambda_1 \geq \frac{\mathbf{e}^\top A \mathbf{e}}{\mathbf{e}^\top \mathbf{e}} > 0.$$

The maximum is achieved for some $\mathbf{x}_1$ with $\|\mathbf{x}_1\| = 1$ and nonnegative coordinates. Thus $\mathbf{x}_1$ must be an eigenvector of A corresponding to λ_1, i.e., $A\mathbf{x}_1 = \lambda_1 \mathbf{x}_1$.

Note that

$$(I_n + A)\mathbf{x}_1 = (1 + \lambda_1)\mathbf{x}_1.$$

Hence

$$(I_n + A)^{n-1}\mathbf{x}_1 = (1 + \lambda_1)^{n-1}\mathbf{x}_1.$$

Since $G(A)$ is connected, Lemma 4.8 implies that $(I_n + A)^{n-1}$ has all positive entries. Since $\mathbf{x}_1$ has at least one positive coordinate, it follows that $(I_n + A)^{n-1}\mathbf{x}_1$ has all positive coordinates. Thus $\mathbf{x}_1 = (1 + \lambda_1)^{-n+1}(I_n + A)^{n-1}\mathbf{x}_1$ has positive coordinates.

Since $|\lambda_i| = |\mathbf{x}_i^\top A \mathbf{x}_i| \le |\mathbf{x}_i|^\top A |\mathbf{x}_i| \le \lambda_1$, it follows that $\lambda_1 \ge |\lambda_i|$ for $i = 2, \ldots, n$.

It remains to show that $\lambda_1 > \lambda_i$ for each $i > 1$. Consider the matrix $B = (b_{ij})_{i,j=1}^n = (I_n + A)^{n-1}$. The eigenvalues of B are $\beta_i = (1 + \lambda_i)^{n-1}$ for $i = 1, \ldots, n$. Clearly,

$$\beta_1 = (1 + \lambda_1)^{n-1} \ge (1 + |\lambda_i|)^{n-1} \ge |\beta_i|.$$

It suffices to show that $\beta_1 > \beta_i$ for $i > 1$.

We repeat the arguments for the maximal eigenvalue β_1 of a real symmetric matrix B with positive entries. Let $\mathbf{y}_1$ be a positive eigenvector of B corresponding to $\max_{\|\mathbf{y}\|=1} \mathbf{y}^\top B \mathbf{y}$. It is known that we can always choose the eigenvector $\mathbf{y}_i$ of B such that $B\mathbf{y}_i = \beta_i \mathbf{y}_i$, $\|\mathbf{y}_i\| = 1$, and $\mathbf{y}_1^\top \mathbf{y}_i = 0$ for all $i > 1$.

Since all coordinates of $\mathbf{y}_1$ are positive, the condition $\mathbf{y}_1^\top \mathbf{y}_i = 0$ implies that $\mathbf{y}_i$ must have both positive and negative coordinates for $i > 1$. As each $b_{pq} > 0$, it follows that

$$|\beta_i| = |\mathbf{y}_i^\top B \mathbf{y}_i| < |\mathbf{y}_i|^\top B |\mathbf{y}_i| \le \beta_1 \quad \text{for } i > 1.$$

Thus $\lambda_1 > \lambda_i$ for each $i > 1$.

Assume now that $G(A)$ is not bipartite. Then $G(A)$ is aperiodic (Problem 2.4-19), which means that there exists an integer N such that for all $m \ge N$, A^m has positive entries. In particular, A^{2N} has positive entries, and thus $G(A^{2N})$ is connected.

The eigenvalues of A^{2N} are $\lambda_1^{2N}, \ldots, \lambda_n^{2N}$, all of which are nonnegative. The eigenvalue λ_1^{2N} is the maximum eigenvalue. Therefore we have

$$\lambda_1^{2N} > \lambda_n^{2N} = |\lambda_n|^{2N},$$

which implies

$$\lambda_1 > |\lambda_n|.$$

Assume that $G(A)$ is bipartite. Then there exists a permutation matrix $Q \in \{0,1\}^{n \times n}$ such that $C = QAQ^\top = \left(\begin{smallmatrix} 0 & A_{12} \\ A_{21} & 0 \end{smallmatrix} \right)$. (Since $C^\top = C$, it follows that $A_{21} = A_{12}^\top$.) Assume that $A_{12} \in \mathbb{R}^{p \times q}$. Clearly, C and A are similar matrices. Therefore C and A have the same eigenvalues. Note that $C(Q\mathbf{x}_1) = \lambda_1(Q\mathbf{x}_1)$. Let $(Q\mathbf{x}_1)^\top = (\mathbf{u}^\top, \mathbf{v}^\top)$, where $\mathbf{u} \in \mathbb{R}^p$ and $\mathbf{v} \in \mathbb{R}^q$ are positive vectors. Then

$$\begin{pmatrix} 0 & A_{12} \\ A_{21} & 0 \end{pmatrix} \begin{pmatrix} \mathbf{u} \\ \mathbf{v} \end{pmatrix} = \lambda_1 \begin{pmatrix} \mathbf{u} \\ \mathbf{v} \end{pmatrix} \Rightarrow \begin{pmatrix} 0 & A_{12} \\ A_{21} & 0 \end{pmatrix} \begin{pmatrix} \mathbf{u} \\ -\mathbf{v} \end{pmatrix} = -\lambda_1 \begin{pmatrix} \mathbf{u} \\ -\mathbf{v} \end{pmatrix}.$$

Hence $-\lambda_n = \lambda_n$. Consider $C^2 = \text{diag}(A_{12}A_{21}, A_{21}A_{12})$. Since $G(C^2)$ is isomorphic to $G(A^2)$, it follows from Problem 2.4-19 that $A_{12}A_{21}$, $A_{21}A_{12}$ are irreducible and aperiodic. Note that

$$A_{12}A_{21}\mathbf{u} = \lambda_1^2 \mathbf{u}, \quad A_{21}A_{12}\mathbf{v} = \lambda_1^2 \mathbf{v}.$$

As the eigenvalues of C^2 are $\lambda_1^2, \ldots, \lambda_n^2$, which are nonnegative, and all the eigenvalues of $A_{12}A_{21}, A_{21}A_{12}$ different from $\lambda_1^2 = \lambda_n^2$ are strictly less than λ_1^2, we deduce that $|\lambda_i| < \lambda_1$ for $i = 2, \ldots, n-1$. □

The arguments in the proof of this theorem that show $\lambda_n = -\lambda_1$ when $G(A)$ is bipartite imply

Corollary 5.17. *Let $A \in S_n(\mathbb{R})$ and assume that $G(A)$ is bipartite. Then the first $\lfloor \frac{n}{2} \rfloor$ eigenvalues of A are nonnegative, and the last $\lfloor \frac{n}{2} \rfloor$ are nonpositive. Furthermore, $\lambda_{n-i+1} = -\lambda_i$ for $i = 1, \ldots, n$.*

Theorem 5.18. *Let $A \in S_n(\mathbb{R}_+)$ and assume that A is irreducible. Let $A\mathbf{x}_1 = \lambda_1\mathbf{x}_1$, where $\mathbf{x}_1$ is a positive vector with norm 1. Then*

$$\|\lambda_1^{-m}A^m - \mathbf{x}_1\mathbf{x}_1^\top\| = r^m \quad \text{for all } m \in \mathbb{N}, \quad \text{where } r := \max_{i \in [2,n]} \frac{|\lambda_i|}{\lambda_1}. \tag{5.7.1}$$

If $G(A)$ is not bipartite, then

$$\lim_{m \to \infty} \lambda_1^{-m}A^m = \mathbf{x}_1\mathbf{x}_1^\top.$$

If $G(A)$ is bipartite, then A^2 decomposes into a direct sum of $A_{12}A_{21} \oplus A_{12}A_{12}$, where each summand is an irreducible aperiodic symmetric matrix. In particular, $\lim_{m \to \infty} \lambda_1^{-2m}A^{2m}$ exists and is a direct sum of two symmetric rank-one matrices, each with a single nonzero eigenvalue 1.

Proof. Recall that $A = X \operatorname{diag}(\lambda_1, \lambda_2, \ldots, \lambda_n)X^\top$ for an orthogonal X whose first column is $\mathbf{x}_1$. Observe next that $\mathbf{x}_1\mathbf{x}_1^\top = X\operatorname{diag}(1, 0, \ldots, 0)X^\top$. Use Theorem 5.15 to deduce

$$\|\lambda_1^{-m}A^m - \mathbf{x}_1^\top\mathbf{x}_1\| = \left\| X\operatorname{diag}\left(1, \frac{\lambda_2^m}{\lambda_1^n}, \ldots, \frac{\lambda_n^m}{\lambda_1^n}\right)X^\top - X\operatorname{diag}(1, 0, \ldots, 0)X^\top \right\|$$

$$= \left\| X\operatorname{diag}\left(0, \frac{\lambda_2^m}{\lambda_1^n}, \ldots, \frac{\lambda_n^m}{\lambda_1^n}\right)X^\top \right\| = r^m.$$

This proves (5.7.1). If $G(A)$ is connected and not bipartite, then $r < 1$. Hence

$$\lim_{k \to \infty} \lambda_1^{-m}A^m = \mathbf{x}_1\mathbf{x}_1^\top = X\operatorname{diag}(1, 0, \ldots, 0)X^\top, \tag{5.7.2}$$

and the theorem follows in this case. The second case follows from the proof of Theorem 5.16. □

Theorem 5.19. *Let $P \in S_n(\mathbb{R}_+)$ be a stochastic matrix, and let $G(P) = ([n], E)$ be the undirected graph induced by P.*
1. *Assume that $G(P)$ is connected. Then:*
   - *$\frac{1}{\sqrt{n}}\mathbf{1}$, where $\mathbf{1} = (1, \ldots, 1)^\top \in \mathbb{R}^n$, is the unique positive P-eigenvector of length one corresponding to the eigenvalue $\lambda_1 = 1$.*

- All other eigenvalues of P are real and lie in the interval $(-1, 1]$. In particular, P has a unique stationary distribution $\mu = \frac{1}{n}(1, \ldots, 1) = \frac{1}{n}\mathbf{1}^\top$, known as the equidistribution.
- (a) If $G(P)$ is not bipartite, then $\lambda_i \in (-1, 1)$ for $i = 2, \ldots, n$, and for $r := \max(|\lambda_2|, |\lambda_n|) < 1$, it follows that

$$\left\| P^m - \mathbf{1}\left(\frac{1}{n}\mathbf{1}^\top\right) \right\| \le r^m, \tag{5.7.3}$$

$$\left\| \mu^{(0)} P^m - \frac{1}{n}\mathbf{1}^\top \right\| \le \|\mu^{(0)}\| r^m \le r^m \quad \text{for all } m \in \mathbb{N}, \tag{5.7.4}$$

$$\lim_{m \to \infty} P^m = \mathbf{1}\left(\frac{1}{n}\mathbf{1}^\top\right), \tag{5.7.5}$$

$$\lim_{m \to \infty} \mu^{(0)} P^m = \mu = \frac{1}{n}\mathbf{1}^\top \quad \text{for all } \mu^{(0)} \in \Pi_n. \tag{5.7.6}$$

- (b) If $G(P)$ is bipartite, then $\lambda_n = -1$, and $\lambda_i \in (-1, 1)$ for $i = 1, \ldots, n-1$. The matrix P^2 decomposes into a direct sum of two stochastic matrices $P_1 \oplus P_2$, where each $G(P_i) = (V_i, E_i)$ is connected and not bipartite. For each $\mu^{(0)} \in \Pi_n$, we have

$$\lim_{m \to \infty} \mu^{(0)} P^{2m} = a\mu_1 \oplus (1-a)\mu_2,$$

where μ_i is the equidistribution corresponding to P_i, and $a \in [0, 1]$ represents $\mathbf{Pr}(X_0 \in V_1)$, i. e., the sum of the coordinates of $\mu^{(0)}$ corresponding to V_1.

2. Assume that $G(P)$ is not connected, and let $G(V_1), \ldots, G(V_k)$ be the connected components of $G(P)$. Then each $P(V_i)$ is a symmetric irreducible stochastic matrix. Furthermore, we have

$$\lim_{m \to \infty} P^{2m} = P_{2,\infty},$$

where $P_{2,\infty}$ is a symmetric stochastic matrix with eigenvalues 1 of multiplicity $m \ge k$, and all other eigenvalues are zero. Each connected component contributes either 1 to m if $G(V_i)$ is not bipartite or 2 if $G(V_i)$ is bipartite. Additionally, $P_{2,\infty}^2 = P_{2,\infty}$. If each connected component is not bipartite, then $m = k$, and

$$\lim_{m \to \infty} P^m = P_{2,\infty} = \bigoplus_{i=1}^{k} \mathbf{1}\mu_i(V),$$

where $\mu_i(V)$ is the unique stationary measure corresponding to $P(V_i)$ for $i = 1, \ldots, k$.

Proof. Using the previous theorems, it suffices to prove the case 1-(a). The inequality $\|P^m - \mathbf{1}(\frac{1}{n}\mathbf{1}^\top)\| \le r^m$ follows from (5.7.1). Furthermore, since $\mu = \mu^{(0)}\mathbf{1}(\frac{1}{n}\mathbf{1}^\top)$, it follows

$$\left\| \mu^{(0)} P^m - \frac{1}{n}\mathbf{1}^\top \right\| = \left\| \mu^{(0)}\left(P^m - \mathbf{1}\left(\frac{1}{n}\mathbf{1}^\top\right)\right) \right\| \le \|\mu^{(0)}\| \cdot \left\| P^m - \mathbf{1}\left(\frac{1}{n}\mathbf{1}^\top\right) \right\| = \|\mu^{(0)}\| r^m.$$

Clearly, $\|\mu^{(0)}\| \leq \sqrt{\|\mu^{(0)}\|_1} = 1$. This proves inequalities (5.7.3) and (5.7.4). The equality parts (5.7.5) and (5.7.6) follow easily. $\qquad\square$

i 5.8 Reversible Markov chains

A Markov chain is said to be *reversible* if it looks the same whether time runs forward or backward. More formally, a Markov chain with transition matrix P and stationary distribution μ is reversible if

$$\mu_i p_{ij} = \mu_j p_{ji} \tag{5.8.1}$$

for all states i and j. This condition is often referred to as the *detailed balance condition*. Intuitively, it means that the long-term rate of transitions from state i to state j is equal to the long-term rate of transitions from state j to state i. To put this into context, we first introduce the concept of a stochastic matrix that is reversible with respect to a distribution. We then return to the concept of a reversible Markov chain.

A stochastic matrix $P = (p_{ij}) \in \mathbb{R}_+^{n \times n}$ is *reversible* with respect to the distribution $\mu = (\mu_1, \ldots, \mu_n) \in \Pi_n$ if

$$\mu_i p_{ij} = \mu_j p_{ji} \quad \text{for all } i, j = 1, \ldots, n. \tag{5.8.2}$$

We begin our discussion of reversible stochastic matrices with the following lemma. For a more detailed treatment, see [1].

Lemma 5.20. *Let $P \in \mathbb{R}_+^{n \times n}$ be a stochastic matrix reversible with respect to $\mu \in \Pi_n$. Then μ is a stationary distribution of P. Define $V = \{i_1, \ldots, i_k\} = supp(\mu) \subset [n]$ as the support of μ, representing the set of vertices in $G(P)$ with positive μ_j values. The submatrix $P(V)$ is stochastic, and $\mu(V) \in \Pi_k$ is its stationary distribution.*

Let $D = diag(\sqrt{\mu_1}, \ldots, \sqrt{\mu_n})$. The submatrix $D(V)$ is a diagonal matrix with positive entries. Moreover, the matrix $T := D(V)PD(V)^{-1}$ is nonnegative, symmetric, and diagonally similar to $P(V)$. Consequently, $P(V)$ possesses only real eigenvalues, each of which is geometrically simple.

Proof. Fix $j \in [1, n]$ in (5.8.2) and sum over $i = 1, \ldots, n$. Since P is stochastic, we have $\mu P = \mu$, implying that μ is a stationary distribution.

Let $V^c = [n] \setminus V$. Suppose first that $V^c \neq \emptyset$. Assuming $i \in V$ and $j \in V^c$ in (5.8.2), we find that $p_{ij} = 0$. Consequently, there are no edges from V to V^c, and $P(V)$ is a stochastic submatrix of P. If $V = [n]$, then trivially $P(V) = P$ is stochastic. Moreover, $\mu(V)$ is the stationary distribution of $P(V)$. To complete the proof, we assume without loss of generality that $V = [n]$. Let $T = DPD^{-1}$. It is straightforward to show that (5.8.2) is equivalent to T being symmetric. Consequently, P is diagonally similar to a symmetric matrix T with nonnegative entries. Let $\lambda_1 \geq \cdots \geq \lambda_n$ be the eigenvalues of T. Then $T = X\mathrm{diag}(\lambda_1, \ldots, \lambda_n)X^\top$, and $P = D^{-1}X\mathrm{diag}(\lambda_1,\ldots,\lambda_n)X^{\top}D$. Therefore $1 = \lambda_1 \geq \cdots \geq \lambda_n$ are the eigenvalues of P, each of which is geometrically simple. $\qquad\square$

Theorem 5.21. *Let $\mu = (\mu_1,\ldots,\mu_n)$ be a probability vector with $\mu_i > 0$ for $i = 1,\ldots,n$. Let $P \in \mathbb{R}_+^{n\times n}$ be a stochastic matrix reversible with respect to μ. Define $D = \mathrm{diag}(\sqrt{\mu_1},\ldots,\sqrt{\mu_n})$. Then $T := DPD^{-1}$ is a nonnegative symmetric matrix diagonally similar to P.*

The associated directed graph $G = G(P) = ([n],E)$ is reversible: $(i,j) \in E$ if and only if $(j,i) \in E$. Let $[n] = \bigcup_{i=1}^{k} V_i$ be the decomposition of G into connected components $G(V_1),\ldots,G(V_k)$. Then $P = \bigoplus_{i=1}^{k} P(V_i)$, where each $P(V_i)$ is an irreducible stochastic matrix reversible with respect to the distribution $\mu_i := \frac{1}{\sum_{j\in V_i}\mu_j}\mu(V_i)$ for $i = 1,\ldots,k$.

Assume now that $G(P)$ is connected and let $1 = \lambda_1 > \lambda_2 \geq \cdots \geq \lambda_n$.

1. *If $G(P)$ is not bipartite (i. e., $-\lambda_n < \lambda_1$), then $r := \max_{i=2,\ldots,n} |\lambda_i| < 1$, and*

$$\|D(P^m - \mathbf{1}\mu)D^{-1}\| \leq r^m, \quad \|(\mu^{(m)} - \mu)D^{-1}\| \leq \|\mu^{(0)}D^{-1}\|r^m, \tag{5.8.3}$$

implying $\lim_{m\to\infty} P^m = \mathbf{1}\mu$.

2. *If $G(P)$ is bipartite with $[n] = V \cup V^c$ and $E \subset V \times V^c \cup V^c \times V$, then $P^2 = Q(V) \oplus Q(V^c)$, where $Q(V)$ and $Q(V^c)$ are irreducible stochastic reversible matrices with respect to $\mu_1 := \frac{1}{\sum_{i\in V}\mu_i}\mu(V)$ and $\mu_2 := \frac{1}{\sum_{i\in V^c}\mu_i}\mu(V^c)$, respectively. Moreover, $G(Q(V))$ and $G(Q(V^c))$ are connected and nonbipartite. Consequently,*

$$\lim_{m\to\infty} P^{2m} = \mathbf{1}\mu_1 \oplus \mathbf{1}\mu_2 \tag{5.8.4}$$

with geometric convergence rate $r := \max_{i=2,\ldots,n-1}\lambda_i^2$, that is,

$$\|D(P^{2m} - \mathbf{1}\mu_1 \oplus \mathbf{1}\mu_2)D^{-1}\| \leq r^m.$$

Proof. The first part of the theorem follows directly from Lemma 5.20. The second part is a consequence of $T := DPD^{-1} \in S_n(\mathbb{R}_+)$, Theorem 5.18, and arguments analogous to those in the proof of Theorem 5.19. $\qquad\square$

Remark 5.22. Note that Theorem 5.4 follows from the above theorem if P is a reversible stochastic matrix with respect to a distribution $\mu = (\mu_1,\ldots,\mu_n)$, where $\mu_i > 0$ for $i = 1,\ldots,n$. Moreover, if $P = P^{\top}$ is stochastic, then P is reversible with respect to the uniform distribution $\mu = \frac{1}{n}\mathbf{1}$. ⚠

Why reversible? We explain the terminology of a reversible Markov chain. Assume that (5.8.2) holds. Then μ is a stationary distribution by condition (5.8.2). Let $X_0, X_1,\ldots$ be the Markov chain associated with P, having stationary distribution μ. Condition (5.8.2) is equivalent to

$$\Pr(X_{m-1} = i, X_m = j) = \Pr(X_{m-1} = j, X_m = i)$$

for all $i, j \in [n]$. This means that we can reverse the random walk in time without altering the probabilities. More generally,

$$\mathbf{Pr}(X_0 = i_0, X_1 = i_1, \ldots, X_m = i_m) = \mathbf{Pr}(X_0 = i_m, X_1 = i_{m-1}, \ldots, X_m = i_0).$$

A class of reversible Markov chains. An important class of examples of time-reversible Markov chains is the random walk on a graph. Considering the degree of each node i, the random walk moves according to the matrix with elements $p_{i,j} = \frac{1}{\deg(i)}$ for each neighbor j of node i and $p_{i,j} = 0$ otherwise. Then it is easy to check that the distribution $\mu(i) = \frac{\deg(i)}{\sum_j \deg(j)}$ satisfies the detailed balance conditions. Thus the random walk is time-reversible, and μ is its stationary distribution.

For example, consider a random walk on the house graph shown in Figure 5.5. The degrees are $(d_1, d_2, d_3, d_4, d_5) = (3, 4, 2, 3, 2)$. So the stationary distribution is $(\mu_1, \mu_2, \mu_3, \mu_4, \mu_5) = (\frac{3}{14}, \frac{4}{14}, \frac{2}{14}, \frac{3}{14}, \frac{2}{14})$.

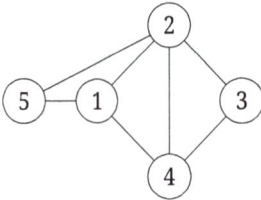

Figure 5.5: Random walk.

ℹ 5.9 Markov chains associated with digraphs

Let $G = ([n], E)$ be a digraph where the outdegree $\deg_{\text{out}}(i) \geq 1$ for each vertex $i = 1. \ldots, n$. With G, we associate the following stochastic matrix:

$$P(G) := P = (p_{ij})_{i,j=1}^n \in \mathbb{R}_+^{n \times n}, \tag{5.9.1}$$

where

$$p_{ij} = \begin{cases} 0 & \text{if } (i, j) \notin E, \\ \frac{1}{\deg_{\text{out}}(i)} & \text{if } (i, j) \in E. \end{cases}$$

On other words, $G(P) = G$, and the probability of transitioning from each vertex i to any vertex j (where i is connected to j) is constant, equal to 1 divided by the number of vertices reachable from i.

Assume that G is reversible, i. e., $(i, j) \in E$ if and only if $(j, i) \in E$. Then $\deg_{\text{out}}(i) = \deg_{\text{in}}(i), i = 1, \ldots, n$, and $\mu_i = \frac{\deg_{\text{out}}(i)}{|E|}, i = 1, \ldots, n$, where $|E| = \sum_{i=1}^n \deg_{\text{out}}(i)$ is the number of directed edges in G. A straightforward calculation shows that P is reversible with respect to the distribution $\mu = (\mu_1, \ldots, \mu_n)$. For this Markov chain, we can use Theorem 5.21.

The issue with $P(G)$ is that even if it is irreducible, it may still be periodic. To avoid this, we can replace the stochastic matrix $G(P)$ with the modified stochastic matrix $P(G, a) := aI_n + (1 - a)P(G)$ for some $a \in (0,1)$. This ensures that the loop $(i, i) \in G(P, a)$ exists for all $a \in (0,1)$. Consequently, the restriction of $P(G, a)$ to any nonempty terminal set $\emptyset \neq V \subset [n]$, where $\{V\}$ is a terminal vertex in the reduced graph G_{rdc}, is an aperiodic stochastic submatrix of $P(G, a)$. (This technique is also employed in the Google search engine!) This guarantees that $\lim_{m \to \infty} P(G, a)^m = Q$, where Q is some special stochastic matrix associated with G independent of $a \in (0,1)$.

In the case $a = 0.5$, $P(G, 0.5)$ represents a *lazy* random walk on the graph G, since with a probability at least 0.5, the particle stays at vertex i at time $m + 1$ if it was at time m at the vertex i.

5.10 Perron–Frobenius theorem

The Perron–Frobenius theorem, established by Oskar Perron in 1907 and Georg Frobenius in 1912, provides a clear characterization of the largest positive eigenvalue and its corresponding positive eigenvector. This theorem is significant because eigenvalue problems involving such matrices appear in various mathematical fields. In probability theory, it is crucial for analyzing the ergodicity of Markov chains. In dynamical systems, it is relevant to the study of subshifts of finite type. See also [16], where the Perron–Frobenius theorem is applied to generalize Kepler's theorem. Edmund Landau was among the first to apply Perron–Frobenius eigenvectors to the problem of ordering players in tournaments.

Theorem 5.23. *Let $0 \neq A \in \mathbb{R}_+^{n \times n}$ be an irreducible matrix, meaning that $G(A)$ is strongly connected. Then the spectral radius $\rho(A)$ is an eigenvalue of A, and it is a simple root of the characteristic polynomial $\det(zI - A)$. The eigenspace corresponding to $\rho(A)$ is spanned by a positive vector $\mathbf{v} > 0$. Let $\mathbf{u} > 0$ be a left positive eigenvector of A corresponding to $\rho(A)$, satisfying $\mathbf{u}^{\top}A = \rho(A)\mathbf{u}^{\top}$. Assume the normalization condition $\mathbf{u}^{\top}\mathbf{v} = 1$. Then any eigenvector $\mathbf{z} \in \mathbb{C}^n$ corresponding to an eigenvalue $\lambda \neq \rho(A)$ of A satisfies $\mathbf{u}^{\top}\mathbf{z} = 0$.*
1. *If $A = (a_{ij})_{i,j=1}^n$ is aperiodic (i. e., $G(A)$ is aperiodic), then*

$$\rho(A) = \lambda_1 > |\lambda_2| \geq \cdots \geq |\lambda_n| \geq 0,$$

and

$$\lim_{m \to \infty} \lambda_1^{-m} A^m = \mathbf{v}\mathbf{u}^{\top}.$$

This convergence is geometric with rate $r := \max_{i \in [2,n]} \frac{|\lambda_i|}{|\lambda_1|}$, meaning that

$$\|\lambda_1^{-m}A^m - \mathbf{v}\mathbf{u}^\top\|_p \leq K_p(A)r^m,$$

where $K_p(A)$ is a constant depending on A and $p \in [1,\infty]$. Furthermore,

$$\rho(A) = \max_{\mathbf{x}=(x_1,\dots,x_n)^\top > 0} \min_{i \in [1,n]} \frac{\sum_{j=1}^n a_{ij}x_j}{x_i} = \min_{\mathbf{x}=(x_1,\dots,x_n)^\top > 0} \max_{i \in [1,n]} \frac{\sum_{j=1}^n a_{ij}x_j}{x_i}. \quad (5.10.1)$$

(This is Wielandt's characterization.)

2. If A is p-periodic with $p \geq 2$ (i. e., $G(A)$ is p-periodic), then A has exactly p algebraically simple eigenvalues on the circle $|z| = \rho(A)$, given by

$$\lambda_i = \rho(A)e^{\frac{(i-1)\sqrt{-1}}{p}} \quad \text{for } i = 1,\dots,p.$$

All other eigenvalues of A satisfy $|\lambda| < \rho(A)$:

$$\rho(A) = \lambda_1 = |\lambda_2| = \cdots = |\lambda_p| > |\lambda_{p+1}| \geq \cdots \geq |\lambda_n|.$$

Moreover, the spectrum of A is invariant under multiplication by $e^{\frac{\sqrt{-1}}{p}}$ (and by any pth root of unity). There exists a permutation matrix P such that $PAP^\top$ is of the form described in (4.2.3). Consequently, the eigenvectors corresponding to λ_i, for $i = 2,\dots,p$ can be derived from $\mathbf{v}$, similarly to the situation described in the proof of Theorem 5.16 when $G(A)$ is bipartite. The matrix A^p decomposes as a direct sum

$$A^p = \oplus_{i=1}^p A_i,$$

where each A_i satisfies the conditions of part 1 of this theorem, and $\rho(A_i) = \rho(A)^p$ for $i = 1,\dots,p$. In particular, $\lambda_1^{-m}A^{pm}$ converges as $m \to \infty$ to a rank p nonnegative matrix, which is a direct sum of rank-one positive matrices.

Remark 5.24. The theorem, (with small variations) is called the Perron–Frobenius theorem. For matrices with positive entries, this result is due to Perron 1907. For nonnegative irreducible matrices, this result is due to Frobenius 1908, 1909, and 1912. The minimax characterization (5.10.1) is due to Wielandt 1950.

Theorem 5.25. Let $A \in \mathbb{R}_+^{n \times n}$. Let $G = G(A)$, and denote by G_{rdc} the reduced graph of G, which is acyclic. Assume that G has k vertices. (The case $k = 1$ corresponds to G being strongly connected.)

Then A is permutationally similar to a nonnegative matrix of the form

$$\begin{pmatrix} A_{11} & A_{12} & \cdots & A_{1k} \\ 0 & A_{22} & \cdots & A_{2k} \\ \vdots & \ddots & \ddots & \vdots \\ 0 & 0 & \cdots & A_{kk} \end{pmatrix}.$$

Each A_{ii} is irreducible. The determinant of $zI - A$ is given by $\det(zI - A) = \prod_{i=1}^{k} \det(zI - A_{ii})$, and the spectrum of A is the union of the spectra of the A_{ii} matrices: spec $(A) = \cup_{i=1}^{k}$ spec (A_{ii}). Hence $\rho(A)$ is an eigenvalue of A. There exists a nonzero nonnegative eigenvector $\mathbf{v}$ corresponding to $\rho(A)$: $A\mathbf{v} = \rho(A)\mathbf{v}$. Furthermore,

$$\lim_{m \to \infty} (\mathbf{1}^{\top} A^m \mathbf{1})^{\frac{1}{m}} = \rho(A) = \limsup_{m \to \infty} (\operatorname{tr} A^m)^{\frac{1}{m}}. \tag{5.10.2}$$

A vertex $\{V_i\}$ in G_{rdc}, corresponding to an irreducible matrix A_{ii}, is called singular if $\rho(A_{ii}) = \rho(A)$ (i. e., $\det(\rho(A)I - A_{ii}) = 0$). Let j be the maximal number of singular vertices in any path in G_{rdc}. (For $j = 1$, this may correspond to a path of length 0.) Then j is the maximal size of a Jordan block in A corresponding to $\rho(A)$. Furthermore, the size of the Jordan block of any eigenvalue λ of A satisfying $|\lambda| = \rho(A)$ does not exceed this value of j.

The proof of Theorem 5.4 follows mainly from Theorem 5.23, Theorem 5.25, and Proposition 5.3.

Proof. Consider $A_k = A + \frac{1}{k}\mathbf{1}\mathbf{1}^{\top}$. Each A_k has positive entries, and $\lim_{k \to \infty} A_k = A$. Using the fact that $\lim_{k \to \infty} \det(zI_n - A_k) = \det(zI_n - A)$, the eigenvalues of A_k, counted with their multiplicities, converge to the eigenvalues of A. Therefore $\lim_{k \to \infty} \rho(A_k) = \rho(A)$, and $\rho(A) \in$ spec (A).

Let $A_k \mathbf{v}_k = \rho(A_k)\mathbf{v}_k$, where $\mathbf{v}_k > 0$ with $\|\mathbf{v}_k\| = 1$. By taking a convergent subsequence $\mathbf{v}_{k_q} \to \mathbf{v}$ we deduce that $\mathbf{v} \geq \mathbf{0}$, $\|\mathbf{v}\| = 1$, and $A\mathbf{v} = \rho(A)\mathbf{v}$. Clearly, spec $(A) = \cup_{i=1}^{k}$ spec (A_{ii}).

Observe that $\|A^m\|_1 \leq \mathbf{1}^{\top} A^m \mathbf{1} \leq n\|A^m\|_1$. Use Theorem 5.15 for $p = 1$ to obtain the first equality in (5.10.2). Since $\mathbf{1}^{\top} A^m \mathbf{1} \geq \operatorname{tr} A^m$, we deduce from the first equality in (5.10.2) that $\rho(A) \geq \limsup_{m \to \infty} (\operatorname{tr} A^m)^{\frac{1}{m}}$. Using the fact that spec $(A) = \cup_{i=1}^{k}$ spec (A_{ii}) and applying the Perron–Frobenius theorem, we deduce the second equality in (5.10.2).

The claim about the maximal size of the Jordan block corresponding to $\rho(A)$ is due to Rothblum [36]. The claim that the maximal size of the Jordan block corresponding to an eigenvalue λ on the circle $|z| = \rho(A)$ is well known to experts. See, for example, [19] for proofs of these claims. □

Remark 5.26. We are now ready to prove Theorem 5.4. The proof follows directly from Theorem 5.23, Theorem 5.25, and Proposition 5.3.

!

5.11 Mean first passage time for irreducible Markov chains

i

Absorbing Markov chains. A Markov chain in which, upon reaching a certain state, it is impossible to leave that state is said to have *absorbing states*. A Markov chain with at least one absorbing state is referred to as an *absorbing Markov chain*. In other words, if we are dealing with a Markov chain with states $S := \{s_1, \ldots, s_n\}$ and transition matrix

$P = (p_{ij})_{i,j=1}^n$, then a state s_i is an absorbing state if, once the system reaches state s_i, it remains there. This is expressed as $p_{ii} = 1$. When a process reaches an absorbing state, we say that it is *absorbed*. In an absorbing Markov chain, a state that is not absorbing is called *nonabsorbing*.

Consider the following transition matrix with 5 states:

$$P = \begin{pmatrix} 1 & 0 & 0 & 0 & 0 \\ \frac{1}{2} & 0 & \frac{1}{2} & 0 & 0 \\ 0 & \frac{1}{3} & \frac{1}{3} & \frac{1}{3} & 0 \\ 0 & 0 & \frac{1}{2} & 0 & \frac{1}{2} \\ 0 & 0 & 0 & 0 & 1 \end{pmatrix}.$$

In this matrix the states s_1 and s_5 are absorbing, whereas the others are nonabsorbing. We can reorder the states so that the nonabsorbing states come first, followed by the absorbing states s_2, s_3, s_4, s_1, s_5. The transition matrix in this order will have the following form:

$$P = \begin{pmatrix} 0 & \frac{1}{2} & 0 & \frac{1}{2} & 0 \\ \frac{1}{3} & \frac{1}{3} & \frac{1}{3} & 0 & 0 \\ 0 & \frac{1}{2} & 0 & 0 & \frac{1}{2} \\ 0 & 0 & 0 & 1 & 0 \\ 0 & 0 & 0 & 0 & 1 \end{pmatrix}.$$

This matrix can be observed to have the following canonical form:

$$P = \begin{pmatrix} Q & R \\ 0 & I_2 \end{pmatrix},$$

where I_2 is the 2×2 identity matrix, 0 is the 2×3 zero matrix, R is a nonzero 3×2 matrix, and Q is a 3×3 matrix.

We can formulate this for any absorbing Markov chain as follows: An absorbing Markov chain can be represented by reordering the states so that the nonabsorbing states come first. If there are r absorbing states and t nonabsorbing states, then the transition matrix has the following canonical form:

$$P = \begin{pmatrix} Q & R \\ 0 & I_r \end{pmatrix},$$

where I_r is the $r \times r$ identity matrix, 0 is the $r \times t$ zero matrix, R is a nonzero $t \times r$ matrix, and Q is a $t \times t$ matrix. The first t states are nonabsorbing, whereas the last r states are absorbing. As stated in Problem 5.13-4, the entry $p_{ij}^{(n)}$ in the matrix P^n represents the probability of reaching state s_j after n steps, starting from state s_i. A typical argument in matrix algebra demonstrates that P^n can be written in the following canonical form:

$$P^n = \begin{pmatrix} Q^n & X_n \\ 0 & I_r \end{pmatrix},$$

where X_n signifies the $t \times r$ matrix located in the upper right corner of P^n.

The form of P^n reveals that the entries of Q^n reflect the probabilities of being in each nonabsorbing state after n steps, given any initial nonabsorbing state.

We prove in Worked-out problem 5.12-2 that if the spectral radius of Q is less than 1, then the probability of being in the nonabsorbing states approaches zero as n increases. Therefore it follows that every entry of Q^n approaches zero in the limit as n approaches infinity (i. e., $Q^n \to 0$).

The fundamental matrix. Our next goal is to define the fundamental matrix of a given absorbing Markov chain. To achieve this, we start with the following:

Proposition 5.27. *For an absorbing Markov chain with $\rho(Q) < 1$, the matrix $I_t - Q$ is invertible with an inverse $N = (n_{ij}) = \sum_{i=0}^{\infty} Q^i$. Furthermore, n_{ij} represents the expected number of times the chain is in state s_j, given that it starts in state s_i. The initial state is counted if $i = j$.*

Proof. Consider the equation $(I_t - Q)\mathbf{x} = 0$, which implies $\mathbf{x} = Q\mathbf{x}$. By iterating this we obtain $\mathbf{x} = Q^n \mathbf{x}$. It follows from Worked-out problem 5.12-2 that $Q^n \to 0$ as $n \to \infty$. Therefore $Q^n \mathbf{x} \to 0$, so $\mathbf{x} = 0$. Hence the inverse $(I_t - Q)^{-1} = N$ exists. Next, observe that

$$(I_t - Q) \sum_{k=0}^{n} Q^k = I_t - Q^{n+1}.$$

Multiplying both sides by N yields

$$\sum_{k=0}^{n} Q^k = N(I_t - Q^{n+1}).$$

As n tends to infinity, we have

$$N = I_t + Q + Q^2 + \cdots.$$

Let s_i and s_j be two nonabsorbing states, and assume throughout the remainder of the proof that i and j are fixed. Define X_m as a random variable that equals 1 if the chain is in state s_j after m steps and 0 otherwise. For each m, this random variable depends on both i and j; however, we omit this dependency for clarity. We have

$$\mathbf{Pr}(X_m = 1) = q_{ij}^{(m)}$$

and

$$\mathbf{Pr}(X_m = 0) = 1 - q_{ij}^{(m)},$$

where $q_{ij}^{(m)}$ is the ijth entry of Q^m. These equations hold for $m = 0$ since $Q^0 = I_t$. Therefore, since X_m is a 0–1 random variable, we have $E(X_m) = q_{ij}^{(m)}$.

The expected number of times the chain is in state s_j in the first n steps, given that it starts in state s_i, is clearly

$$E\left(\sum_{k=0}^{n} X_k\right) = q_{ij}^{(0)} + q_{ij}^{(1)} + \cdots + q_{ij}^{(n)}.$$

Letting n tend to infinity, we obtain

$$E\left(\sum_{k=0}^{\infty} X_k\right) = q_{ij}^{(0)} + q_{ij}^{(1)} + \cdots = n_{ij}.$$

In an absorbing Markov chain P, the matrix $N = (I_t - Q)^{-1}$ is known as the *fundamental matrix* of P. The entry n_{ij} in N indicates the expected number of times the process will visit the nonabsorbing state s_j when starting from the nonabsorbing state s_i. □

Example 5.28. Consider the Markov chain given at the beginning of this section. Then

$$Q = \begin{pmatrix} 0 & \frac{1}{2} & 0 \\ \frac{1}{3} & \frac{1}{3} & \frac{1}{3} \\ 0 & \frac{1}{2} & 0 \end{pmatrix}.$$

Therefore

$$N = (I_3 - Q)^{-1} = \begin{pmatrix} \frac{3}{2} & \frac{3}{2} & \frac{1}{2} \\ 1 & 3 & 1 \\ \frac{1}{2} & \frac{3}{2} & \frac{3}{2} \end{pmatrix}$$

is the fundamental matrix considering the transit states s_2, s_3, and s_4. From the middle row of N we observe that if the process starts in state s_3, then the expected numbers of times it will be in states s_2, s_3, and s_4 before absorption are 1, 3, and 1, respectively.

Mean first passage time (MFPT) and mean recurrence time (MRT). Let P be the transition matrix of an irreducible Markov chain with states $s_1, s_2, \ldots, s_n$. Let $\mu = (\mu_1, \mu_2, \ldots, \mu_n)$ be the unique stationary distribution such that $\mu P = \mu$. By the law of large numbers for Markov chains (Problem 5.13-12), the long-term fraction of time the process spends in state s_i is μ_i. Consequently, regardless of the initial state, the chain will eventually reach state s_i and will in fact visit s_i infinitely often.

Another perspective is as follows: Consider a new Markov chain where s_i is made an absorbing state, i.e., $p_{ii} = 1$. If the process starts in any state other than s_i, then the new chain behaves identically to the original chain up until the first visit to s_i. Since the original chain is irreducible, s_i is reachable from any other state, meaning that the new chain is an absorbing chain with s_i as the sole absorbing state. Hence starting the original chain in any state s_j (with $j \neq i$) guarantees that the chain will eventually reach state s_i.

Let N denote the fundamental matrix for the new chain. As discussed, the entries of N represent the expected numbers of visits to each state before absorption. In the context of the original chain, these entries give the expected number of visits to each state before reaching s_j for the first time. Let t_i represent the expected number of steps before the chain is absorbed, given that it starts in state s_i. Define $\mathbf{t}$ as the column vector where the ith entry is t_i. Then we may argue that

$$\mathbf{t} = N\mathbf{c},$$

where $\mathbf{c}$ is the column vector with all entries equal to 1. If we sum all the entries in the ith row of N, then we obtain the expected number of times the process will be in any of the nonabsorbing states, given that it starts in state s_i. This sum represents the expected time required before absorption. Therefore t_i is the sum of the entries in the ith row of N. Expressing this in matrix form results in $\mathbf{t} = N\mathbf{c}$.

For an irreducible Markov chain, if the process starts in state s_i, then the expected number of steps to reach state s_j for the first time is called the *mean first passage time* (**MFPT**) from s_i to s_j, denoted by m_{ij}. By convention, $m_{ii} = 0$. The mean recurrence time is a quantity that is closely related to the mean first passage time and is defined as follows: Suppose we start in state s_i and look at the duration until we return to s_i for the first time. It is clear that a return will happen; after the first step, we either stay in s_i or move to another state s_j. Since the chain is irreducible, we will eventually return to s_i from any state s_j. If an irreducible Markov chain starts in state s_i, then the expected number of steps to return to s_i for the first time is known as the *mean recurrence time* (**MRT**) for s_i, denoted by r_i.

We aim to investigate the relationship between **MFPT** and **MRT**. Consider the **MFPT** from state s_i to state s_j under the condition that $i \neq j$. The calculation proceeds as follows: we determine the expected number of steps based on the outcome of the first step, multiply this by the probability of that outcome, and sum the results.

If the first step leads directly to state s_j, then the expected number of steps required is 1. Conversely, if the first step transitions to another state s_k, then the expected number of steps becomes $m_{kj} + 1$ to account for the step already taken. This leads to the equation

$$m_{ij} = p_{ij} + \sum_{k \neq j} p_{ik}(m_{kj} + 1). \tag{5.11.1}$$

Since $\sum_k p_{ik} = 1$, we have

$$m_{ij} = 1 + \sum_{k \neq j} p_{ik} m_{kj}. \tag{5.11.2}$$

Similarly, when starting from state s_i, it is clear that at least one step is necessary to return. By evaluating all possible first steps we derive the following expression:

$$r_i = \sum_k p_{ik}(m_{ki} + 1) = 1 + \sum_k p_{ik}m_{ki}. \tag{5.11.3}$$

We now define two matrices associated with **MFPT** and **MRT**. Define $M = (m_{ij})$ with m_{ij} representing the mean first passage time from state s_i to state s_j when $i \neq j$, whereas the diagonal entries are set to 0. The matrix M is referred to as the *mean first passage matrix*. Next, define the matrix D containing all entries as 0, except for the diagonal entries $d_{ii} = r_i$. This matrix D is known as the *mean recurrence matrix*. Let X be the $n \times n$ matrix with all entries equal to 1. Applying equation (5.11.2) for the case where $i \neq j$ and equation (5.11.3) for the case where $i = j$, we arrive at the matrix equation

$$M = PM + X - D, \tag{5.11.4}$$

which means that

$$(I_n - P)M = X - D. \tag{5.11.5}$$

Theorem 5.29. *For an irreducible Markov chain, the mean recurrence time for state s_i is given by $r_i = \frac{1}{\mu_i}$, where μ_i is the ith component of the stationary distribution vector associated with the transition matrix.*

Proof. Multiplying both sides of equation (5.11.5) by the vector μ along with the fact that $\mu(I_n - P) = 0$, we obtain

$$\mu X - \mu D = 0.$$

Here μX results in a row vector with all entries equal to 1, whereas μD is a row vector with the ith entry $\mu_i r_i$. This leads to the equation

$$(1, 1, \ldots, 1) = (\mu_1 r_1, \mu_2 r_2, \ldots, \mu_n r_n),$$

which simplifies to

$$r_i = \frac{1}{\mu_i},$$

thus completing the proof. □

The Kemeny constant of a Markov chain. The Kemeny constant represents the average number of steps a Markov chain requires to reach a randomly chosen state, starting from its stationary distribution, regardless of the initial state. This intriguing property, first identified by John Kemeny in 1960, led to a reward offer for anyone who could provide an intuitive explanation for why this average number remains consistent, irrespective of the chain's starting point.

Let $P = (p_{ij})$ be the transition matrix of an irreducible Markov chain with state space $S = [n]$. The power sequence $\{P^n\}_{n=0}^{\infty}$ is convergent, and we assume that $P \to P^{\infty}$. This limit satisfies $PP^{\infty} = P^{\infty}P = P^{\infty}$. The matrix P^{∞} has identical rows: $(P^{\infty})_{ij} = \mu_j$.

The row vector $\mu = (\mu_1, \mu_2, \ldots, \mu_n)$ is a probability distribution, where $\mu_j \geq 0$, and $\sum_j \mu_j = 1$. This vector μ represents the stationary distribution of the chain, indicating the steady-state probability of being in state j. The row vector μ is uniquely determined, up to scaling by a constant, as the only row vector that satisfies $\sum_i \mu_i p_{ij} = \mu_j$. Consequently, μ is a row eigenvector corresponding to the eigenvalue 1. The associated column eigenvector is a constant vector with $\sum_j p_{ij} = 1$.

Let m_{ij} denote the expected time to transition from state i to state j, with $m_{ii} = 0$. The mean time to reach equilibrium from state i is $m_{i\mu} = \sum_j m_{ij}\mu_j$, representing the expected time to transition from state i to a randomly chosen state j according to the stationary distribution μ.

John Kemeny made the following observation [25], which explains that the Kemeny constant is a constant but does not explain why it is a constant!

Theorem 5.30. $m_{i\mu}$ does not depend on the starting state i.

Note that this common value, denoted by K, is known as the *Kemeny constant* of the chain. In the proof, some knowledge of geometric analysis is assumed. Specifically, we will use the maximum principle (Theorem A.17) for harmonic functions. For further details, the reader is referred to Appendix A.7.

Proof. Note that the function $m_{i\mu}$ is harmonic, meaning that it satisfies the averaging property

$$m_{i\mu} = \sum_j p_{ij}m_{j\mu}.$$

The reason is that moving away a step from state i brings you closer to your destination, except when your destination is i, in which case the journey restarts, returning to i. This happens with probability μ_i, and the expected duration of this repeated journey is $\frac{1}{\mu_i}$ since the mean return time to i equals the reciprocal of the stationary probability μ_i. Therefore

$$m_{i\mu} - 1 + \mu_i \frac{1}{\mu_i} = \sum_j p_{ij}m_{j\mu},$$

which simplifies to

$$m_{i\mu} = \sum_j p_{ij}m_{j\mu}.$$

By applying a consequence of the maximum principle (Theorem A.17 in Appendix A.7), any function f_i that satisfies

$$\sum_j p_{ij} f_j = f_i$$

must be constant. Selecting i to maximize f_i and noting that this maximum is also reached for any j where $p_{ij} > 0$ confirm that $m_{i\mu}$ does not depend on i. □

The application of the maximum principle reveals that the only column eigenvectors of the matrix P associated with the eigenvalue 1 are constant vectors, a detail that was implicitly stated earlier. The preceding discussion illustrates that the mean time from state i to equilibrium is constant. To find this value, we refer to Kemeny's original proof, which involved deriving an explicit formula for $m_{i\mu}$ and demonstrating its independence from i.

Set

$$Z := (I - P^{\infty}) + (P - P^{\infty}) + (P^2 - P^{\infty}) + \cdots = \sum_{n=0}^{\infty} (P^n - P^{\infty})$$

$$= (I - (P - P^{\infty}))^{-1} - P^{\infty}.$$

Assuming $Z = (z_{ij})$, each entry z_{ij} represents the "expected overwhelming number of visits" to j starting from i. Thus

$$\sum_i \mu_i z_{ij} = 0, \quad \text{and} \quad \sum_j z_{ij} = 0.$$

Since z_{ij} evaluates the overwhelming number of visits to state j starting from state i relative to starting in equilibrium, it follows that $z_{jj} \geq z_{ij}$. This is because to accumulate the overwhelming number of visits to j starting from i, we must first reach j. The discrepancy $z_{jj} - z_{ij}$ corresponds to the overwhelming number of visits to j due to starting from i, which is given by $m_{ij}\mu_j$. In equilibrium, this discrepancy represents the expected number of visits to j over an interval of expected length m_{ij}. Hence we obtain the formula

$$m_{ij} = \frac{z_{jj} - z_{ij}}{\mu_j}.$$

Theorem 5.31. *Kemeny's constant K equals the trace of Z:*

$$K = m_{i\mu} = \sum_j z_{jj}.$$

Proof.
$$m_{i\mu} = \sum_j m_{ij}\mu_j = \sum_j z_{jj} - \sum_j z_{ij} = \sum_j z_{jj},$$

invoking the fact that $\sum_j z_{ij} = 0$. □

Remark 5.32. As mentioned earlier, a reward was offered for an intuitively compelling explanation of why the sum above is independent of the initial state i. Peter Doyle proposed the following interpretation of Kemeny's constant. Suppose we have an irreducible Markov chain but we do not know the starting state. We want to begin observing the chain when it can be considered in equilibrium (i. e., as if it were initialized with the stationary distribution μ or after a long period of time). However, we neither know the starting state nor want to wait indefinitely.

Doyle suggested selecting a state according to the stationary distribution μ. Specifically, choose state j with probability μ_j, perhaps using a spinner. Then wait until time t when this state occurs for the first time. We treat t as our starting time and observe the chain from this point forward. Naturally, the probability that we start in state j is μ_j, so we are effectively beginning in equilibrium. Kemeny's constant is the expected value of t, and this value is independent of how the chain was initially started.

5.12 Worked-out problems

1. Given a two-state Markov chain with transition matrix

$$P = \begin{pmatrix} 1-r & r \\ s & 1-s \end{pmatrix},$$

where $r, s \in (0, 1)$, find the stationary distribution μ.

Solution. We know that the stationary distribution μ satisfies the equation $\mu = \mu P$. Thus we have

$$\begin{pmatrix} \mu(1) & \mu(2) \end{pmatrix} = \begin{pmatrix} \mu(1) & \mu(2) \end{pmatrix} \begin{pmatrix} 1-r & r \\ s & 1-s \end{pmatrix},$$

from which we get

$$\begin{cases} \mu(1)(1-r) + \mu(2)s = \mu(1), \\ \mu(1)r + \mu(2)(1-s) = \mu(2). \end{cases}$$

Combining the two equations, we obtain

$$\mu(1)a = \mu(2)\beta.$$

Employing the constraint $\mu(1) + \mu(2) = 1$, we get

$$\mu = \left(\frac{s}{r+s}, \frac{r}{r+s} \right).$$

2. Prove that in an absorbing Markov chain such that $\rho(Q) < 1$, the probability of eventual absorption is equal to 1. In other words, if Q is the nonabsorbing block of the canonical form of our absorbing Markov chain, then show that $\lim_{n \to \infty} Q^n = 0$.

Solution. From each nonabsorbing state s_j, it is possible to reach an absorbing state. Let m_j denote the minimum number of steps required to reach an absorbing state starting from s_j. Define p_j as the probability that starting from s_j, the process does not reach an absorbing state within m_j steps. Clearly, $p_j < 1$. Let m be the maximum of all m_j, and let p be the maximum of all p_j. Consequently, the probability of not being absorbed within m steps is at most p, within $2m$ steps, it is at most p^2, and so on. Since $p < 1$, these probabilities converge to 0.

Moreover, since the probability of not being absorbed within n steps is monotonically decreasing, these probabilities also approach 0. Therefore we conclude that $\lim_{n\to\infty} Q^n = 0$.

5.13 Problems

1. Prove:
   (a) If $P_1, P_2 \in S_n$, then $P_1 P_2 \in S_n$.
   (b) $P \cdot S_n = S_n \cdot P = S_n$ for all $P \in P_n$.
   (c) S_n is a compact set in $\mathbb{R}^{n\times n}$.

2. Let $X_0, X_1, \ldots$ be a homogeneous Markov chain on the state space $[n]$ defined by a primitive stochastic matrix $P = [p_{ij}] \in S_n$. Let $\mu \in \Pi_n$ be the unique stationary distribution of P. Prove that this Markov chain has a unique limiting distribution equal to μ.

3. Prove:
   (a) $\|A\mathbf{x}\|_p \le \|A\|_p \|\mathbf{x}\|_p$ for all $\mathbf{x} \in \mathbb{R}^n$ and $A \in \mathbb{R}^{m\times n}$,
   (b) $\|AB\|_p \le \|A\|_p \|B\|_p$ for all $A \in \mathbb{R}^{m\times n}$ and $B \in \mathbb{R}^{n\times l}$,
   (c) $\|\operatorname{diag}(d_1, \ldots, d_{\min(m,n)})\|_p = \max_{i\in[1,\min(m,n)]} |d_i|$ for any $\operatorname{diag}(d_1, \ldots, d_{\min(m,n)}) \in \mathbb{R}^{m\times n}$,
   (d) $\|A^{l+q}\|_p \le \|A^l\|_p \|A^q\|_p$ for all $l, q \in \mathbb{Z}_+$ and $A \in \mathbb{R}^{n\times n}$.

4. Given a transition matrix $P = (p_{ij})$, prove that the entry $p_{ij}^{(n)}$ of the matrix P^n represents the probability of being in state s_j after n steps, starting from state s_i.

5. Verify that for any matrix $A \in \mathbb{R}^{m\times n}$, we have the following inequality:
$$\frac{1}{\sqrt{n}}\|A\|_1 \le \|A\|_2 \le \sqrt{m}\|A\|_\infty.$$

6. A transition matrix P is said to be *doubly stochastic* if the sum over each column equals one, that is,
$$\sum_i p_{ij} = 1 \quad \text{for all } j.$$

If such a chain is irreducible and aperiodic and consists of $n + 1$ states $0, 1, \ldots, n$, then show that the limiting probabilities are given by

$$\mu_j = \lim_{n\to\infty} p_{ij}^n = \frac{1}{n+1}, \quad j = 0, 1, \ldots, n.$$

7. Let $A \in \mathbb{R}^{n\times n}$ be nonsingular and $n > 1$. Show that the following statements are equivalent:
    (a) A does not have a $k \times (n-k)$ zero submatrix for each $k \in [n-1]$.
    (b) PA is irreducible for each permutation $P \in \mathcal{P}_n$.

8. Let P be an irreducible transition matrix on a finite state space E. Show that a necessary and sufficient condition for P to be aperiodic is the existence of an integer m such that P^m has all its entries positive.

9. Use the Perron–Frobenius theorem to complete the proof of Theorem 5.5.

10. Prove that a matrix $A \in \mathbb{R}_+^{n\times n}$ is irreducible if and only if $cI + A$ is primitive for every scalar $c > 0$.

11. Let $A \in \mathbb{R}_+^{n\times n}$ be an irreducible matrix. Show that A is primitive if and only if one of the following conditions is satisfied:
    (a) $n = 1$, and $A > 0$.
    (b) $n > 1$, and every eigenvalue λ of A other than $\rho(A)$ satisfies $|\lambda| < \rho(A)$.

12. Let $K_{n,j}$ be the proportion of times that an irreducible Markov chain is in state s_j (with probability μ_j) after n steps. Prove that for all $\epsilon > 0$,

$$\lim_{n\to\infty} \mathbf{Pr}\big(|K_{n,j} - \mu_j| \geq \epsilon\big) = 0,$$

    independent of the initial state s_i. This is called the *law of large numbers for irreducible Markov chains*. (See Worked-out problem 1.6-4.)

13. A process transitions among the integers $1, 2, 3, \ldots, n$ starting at 1. At each successive step, it moves to an integer greater than its current position, with equal probability to each of the available larger integers. State n is an absorbing state. Determine the expected number of steps required to reach state n.

6 Symbolic dynamics and walks on graphs

Outline. This chapter initiates a connection between symbolic dynamics and graph theory by interpreting symbolic sequences as walks on graphs, illustrated with a Fibonacci sequence example. We explore *hard-core configurations* on paths, their associated recurrence relations, and an **MCMC** algorithm for this model. Subsequently, we introduce the *Shannon capacity* of a channel represented by a digraph G that contains cycles, investigating its relation to the digraph adjacency matrix spectral radius. The chapter quantifies the number of permissible G-words of a given length. For a distribution $\mu = (\mu_1, \ldots, \mu_n) \in \Pi_n$, we define the entropy

$$H(\mu) := -\sum_{i=1}^{n} \mu_i \log \mu_i$$

and discuss its interpretation as the capacity of a channel transmitting alphabet $[n]$ with letter frequencies μ_i. We introduce the *pressure function*, analyze its analytic properties, and study its correlation with all potentials of permissible configurations. The chapter concludes with a simulation of the pressure gradient.

6.1 From words to walks: connecting information theory and graph theory

View $[n] = \{1, \ldots, n\}$ as an alphabet with n letters. Consider all possible words of this alphabet of length m: $\mathbf{a}_m := (a_1, a_2, \ldots, a_m)$, where $a_i \in [n]$. Each word $\mathbf{a}_m$ can be seen as a walk of length m on the complete digraph $KD_n := ([n], [n] \times [n])$. Specifically, the walk is represented by the sequence $a_1 \to a_2 \to \cdots \to a_m$. The total number of such walks of length m is clearly n^m.

In electrical engineering, each word $\mathbf{a}_m$ represents a transmission of information of length m. This is referred to as an *unconstrained* or *free channel*. The mth capacity of a free channel is $\log n = \frac{\log n^m}{m}$. In information theory, logarithms are typically taken with base 2, so the capacity of the free channel is $\log_2 n$. In mathematics and physics, logarithms are usually taken with the natural base e, denoted simply as $\log x$.

In many cases, there are local restrictions on the form of a word $\mathbf{a}_m$ that can be transmitted. These restrictions can always be translated into the assumption that $\mathbf{a}_m$ is a walk on some graph $G = ([n], E)$.

Consider the following example. Let $G := ([2], \{(1, 2), (2, 1), (2, 2)\})$. This G is a reversible digraph where we can move from state 1 to state 2 and from state 2 to states 1 and 2. In this example, it is common to replace 2 with 0. Thus $\mathbf{a}_m = (a_1, a_2, \ldots, a_m)$ is a signal of length m, consisting of 0s and 1s, such that no two 1s are adjacent. Let l_m be the number of words $\mathbf{a}_m$ that satisfy this restriction. Equivalently, l_m is the number of walks on G of length m. Clearly, $l_1 = 2$ and $l_2 = 3$. We claim that $l_1, l_2, l_3, \ldots$ form the Fibonacci

https://doi.org/10.1515/9783111337388-006

sequence

$$l_m = l_{m-2} + l_{m-1}, \quad m \geq 3. \tag{6.1.1}$$

Consider the word $\mathbf{a}_m = (a_1, a_2, \ldots, a_m)$. First, suppose $a_m = 0$. In this case, $(a_1, \ldots, a_{m-1})$ can be any permissible word of length $m - 1$. Therefore there are l_{m-1} such words of length m that end in $a_m = 0$. Now suppose $a_m = 1$. This implies $a_{m-1} = 0$ because no two 1s can be adjacent. Consequently, $(a_1, \ldots, a_{m-2})$ can be any permissible word of length $m - 2$. Thus there are l_{m-2} such words of length m that end in $a_m = 1$. This proves the Fibonacci-like recurrence relation

$$l_m = l_{m-2} + l_{m-1}, \quad m \geq 3.$$

To find a solution to this recurrence relation, consider the form $l_m = t^m$. Substituting this form into the recurrence relation gives the quadratic equation

$$t^2 = t + 1,$$

which has two solutions,

$$t_1 = \frac{1 + \sqrt{5}}{2} > 0 \quad \text{and} \quad t_2 = \frac{1 - \sqrt{5}}{2} < 0.$$

The number t_1 is known as the *golden ratio*. The general solution to the recurrence relation is

$$l_m = a_1 t_1^m + a_2 t_2^m.$$

Using the initial conditions $l_1 = 2$ and $l_2 = 3$, we find

$$l_m = \left(\frac{5 + 3\sqrt{5}}{10}\right)\left(\frac{1 + \sqrt{5}}{2}\right)^m + \left(\frac{5 - 3\sqrt{5}}{10}\right)\left(\frac{1 - \sqrt{5}}{2}\right)^m, \quad m = 1, 2, \ldots. \tag{6.1.2}$$

Definition 6.1. Let $G = (V, E)$ be an undirected graph. A configuration $\phi : V \to \{0, 1\}$, which assigns either 0 or 1 to each vertex in V, is called *permissible* if $\phi(u) + \phi(v) < 2$ for any two adjacent vertices $u, v \in V$. In other words, no pair of adjacent vertices u and v can both be assigned the value $\phi(u) = \phi(v) = 1$. The set of all permissible configurations $\Phi \subset \{0, 1\}^V$ is known in physics as the model. A permissible configuration $\phi \in \Phi$ is referred to as a configuration.

Let $C_m = ([m], E_m)$ be an undirected path on m vertices: $1 - 2 - \cdots - m$. The edge set E_m is given by $\{(1, 2), (2, 3), \ldots, (m - 1, m)\}$. A *hard-core configuration* on C_m is a word of length m in $\{0, 1\}$ where no two 1s are adjacent. Let Φ_m be the set of all hard-core configurations on C_m. The number of elements in Φ_m is equal to l_m.

Introduce a uniform probability on Φ_m such that $\mathbf{Pr}(\phi) = \frac{1}{l_m}$ for all $\phi \in \Phi_m$. Define the random variable $X_m : \Phi_m \to [m]$ by $X_m(\phi) = \sum_{i=1}^m \phi(i)$, which assigns to each hard-core configuration $\phi \in \Phi_m$ the number of 1s in that configuration.

Let us compute $E(X_m)$. Denote $s_m := \sum_{\phi \in \Phi_m} X_m(\phi)$. Here s_m represents the total number of 1s across all hard-core configurations on C_m. Note that $s_1 = 1$, $s_2 = 2$, $s_3 = 5$, and $s_4 = 10$. The following recursive relation can be shown:

$$s_m = s_{m-1} + s_{m-2} + l_{m-2}, \quad m \ge 3, \tag{6.1.3}$$

where $l_j, j = 1, 2, \ldots$, is the Fibonacci sequence defined in (6.1.1). (See Problem 6.8-1.)

It is known that, under the assumption that l_m satisfies (6.1.1), the general solution of (6.1.3) is of the form

$$s_m = (mb_1 + b_2)\left(\frac{1 + \sqrt{5}}{2}\right)^m + (mb_3 + b_4)\left(\frac{1 - \sqrt{5}}{2}\right)^m, \quad m \ge 1. \tag{6.1.4}$$

The values of b_1, b_2, b_3, and b_4 are completely determined by s_1, s_2, s_3, and s_4. It can be shown that $b_1 > 0$. Thus $E(X_m) = \frac{s_m}{l_m}$. From these results we deduce

$$\lim_{m \to \infty} \frac{\log l_m}{m} = \log\left(\frac{1 + \sqrt{5}}{2}\right), \tag{6.1.5}$$

$$\lim_{m \to \infty} \frac{E(X_m)}{m} = \frac{10b_1}{3 + \sqrt{5}}. \tag{6.1.6}$$

Finally, let us consider another problem related to the number of configurations with a specific number of 1s. For a nonnegative integer t, define $f(t, m) := l_m \mathbf{Pr}(X_m = t)$. This represents the total number of configurations in Φ_m that have exactly t 1s. Note that for $t > \lceil \frac{m}{2} \rceil$, $f(t, m) = 0$, since no configuration can have more than half the chain filled with 1s.

Fix $p \in [0, 0.5)$ and consider the sequence of integers $t_m \in \mathbb{Z}_+$, $m \in \mathbb{N}$, such that $\lim_{m \to \infty} \frac{t_m}{m}$.

The question is: Does the sequence $\frac{\log f(t_m, m)}{m}$, where m ranges over all natural numbers, converge to a specific function $h(p)$? This function would depend only on the chosen value p and not on the particular sequence t_m we pick.

We will prove that the answer to this question is *positive*. In the next sections, we will explore generalizations of these problems in the context of the hard-core model on chain C_m as the chain length tends to infinity.

i 6.2 MCMC algorithm for the hard-core model

Given a probability distribution μ on $S = \{s_1, \ldots, s_n\}$, how can we simulate a random object with distribution μ? Consider a graph $G = (V, E)$ with vertex set $V = [n]$ and edge set $E = \{e_1, \ldots, e_l\}$. As mentioned previously, a hard-core model on G involves assigning a value of 0 or 1 to each vertex, ensuring that no adjacent vertices are both assigned 1. These assignments, or configurations, can be represented as elements of the set $\{0,1\}^V$. In other words, a sequence $(a_1, \ldots, a_n)$, where $a_1, \ldots, a_n \in \{0,1\}$, gives a map $\phi : V \to \{0,1\}$ with values $\phi(v_i) = a_i$ for $i \in [n]$. A configuration is *feasible* if no two adjacent vertices are both assigned 1.

We define the probability measure $\mathbb{M}_G$ on $\{0,1\}^V$ by assigning equal probability to each feasible configuration. Formally, for $\Psi \in \{0,1\}^V$,

$$\mathbb{M}_G(\Psi) = \begin{cases} \frac{1}{l_G} & \text{if } \Psi \text{ is feasible,} \\ 0 & \text{otherwise,} \end{cases} \tag{6.2.1}$$

where l_G is the total number of feasible configurations for G.

Originally introduced in statistical physics to model a gas with particles of nonnegligible size, the hard-core model (with G as a three-dimensional grid) captures the behavior of particles that cannot overlap. Here 1 represents a particle, and 0 represents the empty space. The model has also found applications in telecommunications to represent scenarios where an active node disables its neighbors.

A natural question is: What is the expected number of occupied vertices in a random configuration under $\mathbb{M}_G$? Let $n(\Psi)$ denote the number of occupied vertices in configuration Ψ, and let X be a random configuration chosen according to $\mathbb{M}_G$. Then this expected value is given by

$$E[n(X)] = \sum_{\Psi \in \{0,1\}^V} n(\Psi) \mathbf{Pr}_G(\Psi) = \frac{1}{l_G} \sum_{\Psi \in \{0,1\}^V} n(\Psi) \mathbf{I}_{\{\Psi \text{ is feasible}\}},$$

where $\mathbf{I}_{\{\Psi \text{ is feasible}\}}$ is the *indicator function* of $\{\Psi \text{ is feasible}\}$, meaning that

$$\mathbf{I}_{\{\Psi \text{ is feasible}\}} = \begin{cases} 1 & \text{if } \Psi \text{ is feasible,} \\ 0 & \text{otherwise} \end{cases}$$

Evaluating this sum directly can be computationally prohibitive, especially for larger graphs, as the number of configurations grows exponentially with the graph size. Although many terms in the sum are zero, the number of nonzero terms also grows exponentially. Additionally, calculating l_G itself is computationally challenging.

In such cases, simulation provides a practical alternative. By generating numerous random configurations according to $\mathbb{M}_G$ and averaging the number of 1s in each we can approximate $E(n(X))$. The law of large numbers (Worked-out problem 1.6-4) guarantees that this approximation converges to the true value as the number of simulations increases, allowing for the construction of confidence intervals using standard statistical methods.

With this example in mind, let us discuss how to simulate a random variable X distributed according to a given probability distribution μ on a state space S. In principle, this is straightforward: enumerate the elements of S as $s_1, \ldots, s_n$ and let $X = \psi(U)$, where U is a uniform $[0,1]$ random variable, and the function $\psi : [0,1] \to S$ is defined as follows:

$$\psi(x) = \begin{cases} s_1 & \text{if } x \in [0, \mu(s_1)), \\ s_2 & \text{if } x \in [\mu(s_1), \mu(s_1) + \mu(s_2)), \\ \vdots \\ s_i & \text{if } x \in [\sum_{j=1}^{i-1} \mu(s_j), \sum_{j=1}^{i} \mu(s_j)), \\ \vdots \\ s_n & \text{if } x \in [\sum_{j=1}^{n-1} \mu(s_j), 1]. \end{cases}$$

We may argue that this gives X the desired distribution μ.

However, this approach is impractical for large state spaces S. This is where the Markov chain Monte Carlo (**MCMC**) method becomes useful [29]. Originating in physics in the 1950s, the method saw significant growth in other fields, particularly in image analysis during the 1980s and in Bayesian statistics in the 1990s.

The concept is as follows: Suppose we can construct an irreducible and aperiodic Markov chain $(X_0, X_1, \ldots)$, with a stationary distribution μ. If we start the chain with an arbitrary initial distribution (e. g., starting from a fixed state), Theorem 5.4 guarantees that the distribution of the chain at time n will converge to μ as n approaches infinity. Therefore, if we run the chain for a sufficiently long time n, then the distribution of X_n will closely approximate μ. Although this is an approximation, it can be made as accurate as desired by selecting n large enough.

We might wonder how constructing a Markov chain to achieve this property could be easier than directly constructing a random variable with distribution μ. To address this, we proceed with an example.

Example 6.2. Let us consider the hard-core model in the previous discussion on a graph $G = (V, E)$ with $V = [n]$. To develop an **MCMC** algorithm for this model, we aim to construct a Markov chain whose state space $\mathcal{S}$ is the set of feasible configurations for G, i. e.,

$$\mathcal{S} = \{\Psi \in \{0,1\}^V : \Psi \text{ is feasible}\}.$$

Additionally, we want the Markov chain to be irreducible and aperiodic and to have the stationary distribution $\mathbb{M}_G$ given by (6.2.1).

A Markov chain $(X_0, X_1, \ldots)$ with the desired properties can be obtained using the following transition mechanism. At each integer time $n + 1$, proceed as follows:
- Pick a vertex $v \in V$ at random (uniformly).
- Toss a fair coin.

- If the coin comes up heads, and all neighbors of v take the value 0 in X_n, then let $X_{n+1}(v) = 1$; otherwise, let $X_{n+1}(v) = 0$.
- For all vertices w other than v, leave the value at w unchanged, i. e., let $X_{n+1}(w) = X_n(w)$.

It is not difficult to verify that this Markov chain is irreducible and aperiodic (verify!). Hence it remains to show that $\mathbb{M}_G$ is a stationary distribution for the chain. By Lemma 5.20 it suffices to show that $\mathbb{M}_G$ is reversible. Letting $P_{\Psi,\Psi'}$ denote the transition probability from state Ψ to state Ψ' (with the transition mechanism as described above), we need to check that

$$\mathbb{M}_G(\Psi)P_{\Psi,\Psi'} = \mathbb{M}_G(\Psi')P_{\Psi',\Psi}$$

for all feasible configurations Ψ and Ψ'. Let $\mathbf{d} = \mathbf{dis}(\Psi, \Psi')$ denote the number of vertices in which Ψ and Ψ' differ, and consider the cases $\mathbf{d} = 0$, $\mathbf{d} = 1$, and $\mathbf{d} \geq 2$ separately. Firstly, the case $\mathbf{d} = 0$ means that $\Psi = \Psi'$, in which case the relation is trivially true. Secondly, the case $\mathbf{d} \geq 2$ is almost as trivial, because the chain never changes the values at more than one vertex at a time, making both sides of the equation equal to 0. Finally, consider the case $\mathbf{d} = 1$, where Ψ and Ψ' differ at exactly one vertex v. Then all neighbors of v must take the value 0 in both Ψ and Ψ', since otherwise one of the configurations would not be feasible. We therefore have

$$\mathbb{M}_G(\Psi)P_{\Psi,\Psi'} = \frac{1}{l_G}\frac{1}{2n} = \mathbb{M}_G(\Psi')P_{\Psi',\Psi},$$

and the relation is verified (recall that n is the number of vertices). Hence the chain has $\mathbb{M}_G$ as a reversible (and therefore stationary) distribution.

We can now simulate this Markov chain. A convenient choice of update function ϕ is to split the unit interval $[0,1]$ into $2n$ subintervals of equal length $\frac{1}{2n}$, representing the choices $(v_1, \text{heads}), (v_1, \text{tails}), (v_2, \text{heads}), \ldots, (v_n, \text{tails})$ in the above description of the transition mechanism. If we run the chain for a long time t, starting with an arbitrary feasible initial configuration such as the "all 0s" configuration, and output X_t, then we get a random configuration whose distribution is approximately $\mathbb{M}_G$.

The algorithm described is a typical example of an **MCMC** method in several ways. Although it suffices for the chain to have the desired distribution as its stationary distribution, we have identified a chain with the stronger property of being reversible. This characteristic is common to the vast majority of known **MCMC** algorithms. The reason is that in most nontrivial cases, the simplest way to construct a chain with given stationary distribution μ is to ensure the reversibility condition (5.8.1).

Historical note. In the years following World War II, Los Alamos became a hub of innovation in applied mathematics and theoretical physics, largely driven by the urgent need to develop nuclear weapons. A particularly challenging problem was understanding the behavior of massive collections of atomic particles, on the order of 10^{23}. The laws governing these particles—thermodynamics, statistical physics, and quantum mechanics—are inherently probabilistic and complex, making traditional analytical methods insufficient for the detailed analysis required. In response to this challenge, a new approach emerged: instead of relying on closed-form, analytical solutions, researchers began simulating the system behavior to estimate solutions. However, generating these simulations was not straightforward. Prior to the late 1940s, no machine could efficiently perform the large-scale random simulations needed. However, with the advent of the ENIAC (Electronic Numerical Integrator and Computer) at the University of Pennsylvania, available to researchers at Los Alamos National Laboratory (Figure 6.1) after the war, this became possible. In 1947, von Neumann and his colleagues were developing methods to estimate neutron diffusion and multiplication rates in fission devices, such as nuclear bombs. Following Stanisław Ulam's suggestion (Figure 6.2), von Neumann proposed a straightforward plan: generate a relatively large number of "virtual" neutrons and use the computer to randomly simulate their evolution through the fissionable material. Upon completion, the number of remaining neutrons would be counted to estimate the desired rates. In modern terms the scale was quite modest; a simulation involving just 100 neutrons with 100 collisions each required approximately five hours of computing time on the ENIAC. Nevertheless, the utility of this approach was immediately evident. From this point onward, randomized simulations, later known as *Monte Carlo methods*, became an essential technique in physics. It is worth noting that Nicholas Metropolis was credited with coining the term "Monte Carlo methods." Consult with [22] for further discussion on Monte Carlo methods.

Figure 6.1: Los Alamos National Laboratory (1943). Photo credit: Wikipedia, Aerial view. Source: https://en.wikipedia.org/wiki/Los_Alamos_National_Laboratory.

Figure 6.2: Stanisław Ulam (1909–1984). Photo credit: Wikipedia, Ulam at Los Alamos. Source: https://en. wikipedia.org/wiki/Stanis%C5%82aw_Ulam.

6.3 Shannon capacity of a channel

Consider the alphabet of n letters denoted by $[n]$. Let $G = ([n], E)$ be a digraph that contains a cycle. A word $\mathbf{a}_m = (a_1, a_2, \ldots, a_m)$ is called *permissible*, or *G-permissible*, if the letter j can follow the letter i, i. e., $(i, j) \in G$. In other words, the word $\mathbf{a}_m$ describes a walk on G, where a_q is the vertex location on the G at time q. (Time starts at $q = 1$ and then is equal to $2, 3, \ldots$) Let $W(m)$ be the set of all permissible words of length m. The assumption that G contains a cycle is equivalent to the assumption that $W(m) \neq \emptyset$ for all $m \in \mathbb{N}$.

Theorem 6.3. *Let $n \in \mathbb{N}$ and consider the alphabet with n letters denoted by $[n]$. Let $G = ([n], E)$ be a directed graph that contains a cycle. Define $A = A(G)$ as the adjacency matrix of G, and let $\rho(A) = \rho(G)$ be the spectral radius of A (and hence of G).*

Let $W(m)$ represent the set of G-permissible words of length m, and denote $l_m = |W(m)|$. Then $l_m = \mathbf{1}^T A^m \mathbf{1}$, where $\mathbf{1}$ is the vector of all 1s. The sequence $\log l_m$ is a nonnegative subadditive sequence.

Thus the limit $h(G) := \lim_{m \to \infty} \frac{\log l_m}{m}$ exists and is called the Shannon capacity of the channel represented by G. Moreover, $h(G) = \log \rho(G)$, and $h(G) \leq \frac{\log l_m}{m}$ for all $m \in \mathbb{N}$.

Proof. Invoking Lemma 4.4, if $A^m = (a_{ij}^{(m)})_{i,j=1}^n$, then $a_{ij}^{(m)}$ represents the number of walks from vertex i to vertex j in m steps. Hence $l_m = \sum_{i,j=1}^n a_{ij}^{(m)} = \mathbf{1}^T A^m \mathbf{1}$. Since G contains a cycle, it follows that $l_m \geq 1$.

Next, observe that $l_{p+q} \leq l_p l_q$ for all $p, q \in \mathbb{N}$. To see why, consider a word of length $p + q$, $a_{p+q} = (a_1, \ldots, a_{p+q}) \in W(p + q)$, which can be viewed as a concatenation of two words, $\mathbf{b} = (a_1, \ldots, a_p) \in W(p)$ and $\mathbf{c} = (a_{p+1}, \ldots, a_{p+q}) \in W(q)$. In general, a concatenation of two words $\mathbf{b} = (b_1, \ldots, b_p) \in W(p)$ and $\mathbf{c} = (c_1, \ldots, c_q)$ to form $\mathbf{a} = (b_1, \ldots, b_p, c_1, \ldots, c_q)$ does not necessarily belong to $W(p + q)$. This depends on whether $(b_p, c_1) \in E$ or not. Hence $l_{p+q} \leq l_p l_q$.

Thus $\log l_m$, $m \in \mathbb{N}$, is a nonnegative subadditive sequence. By Lemma 5.13 the sequence $\frac{\log l_m}{m}$ converges to a (nonnegative) limit denoted by $h(G)$. Furthermore, $h(G) \le \frac{\log l_m}{m}$ for all $m \in \mathbb{N}$. The equality $l_m = \mathbf{1}^\top A^m \mathbf{1}$ and (5.10.2) imply that $h(G) = \log \rho(A)$. $\square$

Infinite words. Denote by $[n]^{\mathbb{N}}$ the set of all mappings $\mathbf{a} : \mathbb{N} \to [n]$, that is, $\mathbf{a}$ can be identified with an infinite sequence $\mathbf{a} = (a_1, a_2, \ldots)$ where $a_i \in [n]$ for all $i \in \mathbb{N}$. We can view $\mathbf{a}$ as an infinite word on the alphabet $[n]$. An infinite word $\mathbf{a}$ is called *periodic* if there exists $q \in \mathbb{N}$ such that $a_{i+q} = a_i$ for all $i \in \mathbb{N}$. A word with this property is said to be *q-periodic*. The smallest such q is called the *period* of $\mathbf{a}$. Thus if the period of $\mathbf{a}$ is p, then $\mathbf{a}$ is q-periodic if and only if p divides q.

Let $G = ([n], E)$ be a digraph that contains a loop. Denote

$$[n]^{\mathbb{N}}(G) := \{\mathbf{a} = (a_1, \ldots) \in [n]^{\mathbb{N}} : (a_i, a_{i+1}) \in E, i = 1, 2, \ldots\}.$$

Thus $\mathbf{a} \in [n]^{\mathbb{N}}(G)$ can be considered an infinite permissible word or an infinite walk on G. If $\mathbf{a} \in [n]^{\mathbb{N}}(G)$ is q-periodic, then $\mathbf{a}_q := (a_1, \ldots, a_q) \in W(q)$, and $(a_q, a_1) \in E$. The set of all q-permissible words $\mathbf{a}_q \mathbf{n} W(q)$ satisfying the condition $(a_q, a_1) \in E$ is denoted by $W_{\mathrm{per}}(q)$. It is possible for $W_{\mathrm{per}}(q)$ to be empty for some q, for instance, if G consists of a single cycle. Thus $\mathbf{a}_q \in W_{\mathrm{per}}(q)$ if and only if $\mathbf{a}_q = (a_1, \ldots, a_q)$ can be extended to a q-periodic word $\mathbf{a} \in [n]^{\mathbb{N}}(G)$. Equivalently, $W_{\mathrm{per}}(q)$ can be identified with all $\mathbf{a}_{q+1} = (a_1, \ldots, a_{q+1}) \in W(q+1)$ where $a_{q+1} = a_1$. Thus $\mathbf{a}_{q+1}$ corresponds to a *closed walk* on G of length q.

Proposition 6.4. *Let $G = ([n], E)$ be a directed graph containing a cycle. Let $A = A(G)$ be its adjacency matrix. Denote by $\mathcal{W}(q)$ the set of all G-permissible words of length q and by $\mathcal{W}_{per}(q)$ the set of all projections onto the first q coordinates of q-periodic words in $[n]^{\mathbb{N}}(G)$. Set $l_q := |W(q)|$ and $l_{q,per} := |W_{per}(q)|$. Then $l_q \ge l_{q,per}$, and*

$$l_{q,\mathrm{per}} = \operatorname{tr} A^q, \quad q \in \mathbb{N}. \tag{6.3.1}$$

Furthermore,

$$\limsup_{q \to \infty} \frac{\log l_{q,\mathrm{per}}}{q} = h(G) = \log \rho(G).$$

Proof. Recall that if $A^q = (a_{ij}^{(q)})_{i,j=1}^n$, then $a_{ij}^{(q)}$ is the number of walks from i to j in m steps (Lemma 4.4). Hence $\operatorname{tr} A^q = \sum_{i=1}^n a_{ii}^{(q)}$ is the number of closed walks on G of length q, i. e., $\operatorname{tr} A^q = l_{q,\mathrm{per}}$. The second part of (6.3.1) follows from the second part of (5.10.2).

Note that in the second part of (6.3.1), we cannot generally replace lim sup with lim. Indeed, assume that G is strongly connected and $2 \le p$-periodic. If a_{q+1} is a closed walk on G of length q, then p divides q. Thus, if p does not divide q, then $\operatorname{tr} A^q = 0$. This shows that in this case we cannot replace lim sup with lim in (6.3.1).

In the case where G is strongly connected and aperiodic, we can replace lim sup with lim in (6.3.1) (refer to Part 1 of Theorem 5.23). $\square$

6.4 Entropy of Markov chains

Let $f(x) = -x \log x$ $x \in \mathbb{R}_+$. Then $f(0) = f(1) = 0$, $f(x) > 0$ for $x \in (0,1)$, and $f(x) < 0$ for $x > 1$. Note that $f''(x) = -\frac{1}{x} < 0$ for $x > 0$. Hence f is a concave continuous function on $\mathbb{R}_+$.

For a distribution $\mu = (\mu_1, \ldots, \mu_n) \in \Pi_n$, the quantity $H(\mu) := -\sum_{i=1}^{n} \mu_i \log \mu_i$ is called the *entropy* of μ. The function H was introduced by Boltzmann (1844–1906) in the Boltzmann H-theorem, in his work on statistical mechanics. (Not to be confused with the entropy concept in thermodynamics, introduced by Rudolf Clausius in 1850 and used by Boltzmann.)

In probability theory, the entropy $H(\mu)$ quantifies the uncertainty associated with the outcome of a random variable X with probability mass function given by $\mathbf{Pr}(X = i) = \mu_i$ for $i = 1, \ldots, n$. It can be shown that $H(\mu) \leq H(\frac{1}{n}\mathbf{1}) = \log n$, with equality if and only if μ corresponds to the uniform distribution, i. e., $\mu_i = \frac{1}{n}$ for all i. Conversely, if $\mu = \mathbf{e}_i = (\delta_{i1}, \ldots, \delta_{in})$, then $H(\mathbf{e}_i) = 0$, indicating complete certainty that $X = i$.

The entropy $H(\mu)$ measures the capacity of a channel transmitting the alphabet $[n]$ such that the frequency of each letter i is μ_i for $i = 1, \ldots, n$. To see this, consider all words $\mathbf{a}_m = (a_1, \ldots, a_m) \in [n]^m$, where each letter i appears $m_i \in \mathbb{N}$ times with $m_1 + \cdots + m_n = m$. The total number of such words is given by

$$t(\mathbf{m}) = \frac{m!}{m_1! \ldots m_n!},$$

where $\mathbf{m} = (m_1, \ldots, m_n) \in \mathbb{Z}_+^n$. Let $p_i = \frac{m_i}{m}$ for $i = 1, \ldots, n$, and let $\mathbf{p} = (p_1, \ldots, p_n) \in \Pi_n$. Recall Stirling's formula

$$k! \approx \sqrt{2\pi k}\left(\frac{k}{e}\right)^k \quad \text{as } \mathbb{N} \ni k \to \infty.$$

Using Stirling's formula, a straightforward calculation shows that

$$\lim_{m \to \infty, \frac{m_i}{m} \to \mu_i, i \in [n]} \frac{\log t(\mathbf{m})}{m} = H(\mu). \tag{6.4.1}$$

Indeed,

$$\log t(\mathbf{m}) = \log m! - \sum_{i=1}^{n} \log m_i!$$

$$\approx m \log m - \sum_{i=1}^{n} m_i \log m_i + O(\log m)$$

$$= -\sum_{i=1}^{n} m_i \log \frac{m_i}{m} + O(\log m)$$

$$= -m \sum_{i=1}^{n} p_i \log p_i + O(\log m).$$

Divide the above equation by m and let $m \to \infty$ to obtain (6.4.1).

Consider the Markov chain defined by the stochastic matrix $P = (p_{ij})_{i,j=1}^{n}$. Assume that $\mu^{(1)} = (\mu_1^{(1)}, \dots, \mu_n^{(1)}) \in \Pi$ is a given initial distribution. Let $X_m \in [n]$ denote the position of a particle performing a random walk on the graph $G = G(P) = ([n], E)$. We can interpret this random walk as occurring on the complete directed graph $KD_n = ([n], E)$, where the transition probabilities are given by

$$\mathbf{Pr}(X_m = i_m \mid X_{m-1} = i_{m-1}, X_{m-2} = i_{m-2}, \dots, X_1 = i_1) = p_{i_{m-1}i_m}.$$

Assume that $\mathbf{Pr}(X_1 = i) = \mu_i^{(1)}$ for $i = 1, \dots, n$. Then

$$\mathbf{Pr}(X_1 = i_1, X_2 = i_2, \dots, X_{m-1} = i_{m-1}, X_m = i_m) = \mu_{i_1}^{(1)} p_{i_1 i_2} \cdots p_{i_{m-1} i_m}. \tag{6.4.2}$$

This indicates that the probability of the walk $(i_1, i_2, \dots, i_{m-1}, i_m)$ is $\mu_{i_1}^{(1)} \prod_{j=1}^{m-1} p_{i_j i_{j+1}}$.

We now consider a channel with the following property: Given that a letter a_{m-1} was transmitted at time $m - 1$, the probability of the next transmitted letter a_m is $p_{a_{m-1}a_m}$. Assuming that the probability of the first letter being a_1 is $\mu_{a_1}^{(1)}$, the probability of the word $\mathbf{a}_m = (i_1, \dots, i_m)$ is given by the right-hand side of equation (6.4.2). The *normalized entropy of all words of length m* is

$$-\frac{1}{m} \sum_{i_1, \dots, i_m \in [n]} \mu_{i_1}^{(1)} \prod_{j=1}^{m-1} p_{i_j i_{j+1}} \log\left(\mu_{i_1}^{(1)} \prod_{k=1}^{m-1} p_{i_k i_{k+1}} \right)$$

$$= -\frac{1}{m} \sum_{i_1, \dots, i_m \in [n]} \mu_{i_1}^{(1)} \prod_{j=1}^{m-1} p_{i_j i_{j+1}} \left(\log \mu_{i_1}^{(1)} + \sum_{k=1}^{m-1} \log p_{i_k i_{k+1}} \right)$$

$$= -\frac{1}{m} \sum_{i_1, \dots, i_m \in [n]} \mu_{i_1}^{(1)} \prod_{j=1}^{m-1} p_{i_j i_{j+1}} \log \mu_{i_1}^{(1)}$$

$$- \frac{1}{m} \sum_{k=1}^{m-1} \left(\sum_{i_1, i_{k-1}, i_{k+1}, \dots, i_m \in [n]} \mu_{i_1}^{(1)} \prod_{j=1}^{m-1} p_{i_j i_{j+1}} \right) \log p_{i_k i_{k+1}}$$

$$= -\frac{1}{m} \sum_{i=1}^{n} \mu_i^{(1)} \log \mu_i^{(1)} - \sum_{i,j=1}^{n} \frac{1}{m} \sum_{k=1}^{m-1} (\mu^{(1)} P^{k-1})_i p_{ij} \log p_{ij}.$$

The last equality is derived as follows. In the expression involving $\log \mu_{i_1}^{(1)}$, sum over the indices $i_m, i_{m-1}, \dots, i_2$ and utilize the stochasticity of P, i. e., $\sum_{i_{j+1}=1}^{n} p_{i_j i_{j+1}} = 1$ for $j = m - 1, \dots, 1$. This yields the first expression in the last equality. To derive the second part, fix $k \in [1, n]$. Then sum over the indices $i_1, \dots, i_{k-1}$ and employ the definition of matrix multiplication to obtain the expression $(\mu^{(1)} P^{k-1})_{i_k}$. Subsequently, sum over the indices $i_m, \ldots, i_{k+2}$ to arrive at the expression $(\mu^{(1)} P^{k-1})_{i_k} p_{i_k i_{k+1}} \log p_{i_k i_{k+1}}$ and apply the stochasticity of P once more. Next, sum over i_k, i_{k+1} to obtain the second part of the last equality.

Now let $m \to \infty$ and use Problem 4.4-5 to deduce that

$$\lim_{m \to \infty} \frac{1}{m} \sum_{k=1}^{m-1} \mu^{(1)} P^{k-1} = \mu \in \Pi_n, \quad \mu = P\mu. \tag{6.4.3}$$

Note that any stationary distribution μ of P can be obtained this way; e. g., assume that $\mu^{(1)} = \mu$. Hence the *entropy* of the given Markov chain is given by

$$h(P, \mu) := - \sum_{i,j=1}^{n} \mu_i p_{ij} \log p_{ij}. \tag{6.4.4}$$

If P has a unique stationary distribution, e. g., P is irreducible, then $h(P) := h(P, \mu)$ is uniquely defined.

Theorem 6.5. *Let $G = ([n], E)$ be a digraph such that the outdegree of every vertex in G is positive. Let $A = (a_{ij})_{i,j=1}^n \in \{0,1\}^{n \times n}$ be the adjacency matrix of G, and denote by $\rho(A)$ the spectral radius of A. Let $P \in [0,1]^{n \times n}$ be a stochastic matrix such that $G(P)$ is a subgraph of G. Let $\mu \in \Pi_n$ be a stationary distribution of P, meaning that $\mu P = \mu$. Then $h(P, \mu) \leq h(G) = \log \rho(A)$. Furthermore, there exist a stochastic matrix P (with $G(P) \subseteq G$) and a stationary distribution μ of P such that $h(P, \mu) = h(G)$. If G is strongly connected, then P is a unique irreducible stochastic matrix given by*

$$P = (p_{ij})_{i,j=1}^n, \quad p_{ij} = \frac{a_{ij} u_j}{\rho(A) u_i}, \quad i, j = 1, \ldots, n, \quad A\mathbf{u} = \rho(A)\mathbf{u}, \quad \mathbf{u} = (u_1, \ldots, u_n)^\top > 0. \tag{6.4.5}$$

The stationary distribution $\mu = (\mu_1, \ldots, \mu_n)$ is unique and given by $\mu_i = u_i v_i$ for $i = 1, \ldots, n$, where

$$A^\top \mathbf{v} = \rho(A)\mathbf{v}, \quad \mathbf{v} = (v_1, \ldots, v_n)^\top > 0, \quad \mathbf{v}^\top \mathbf{u} = 1. \tag{6.4.6}$$

Proof. The inequality $h(P, \mu) \leq h(G)$ follows from the fact that the Markov chain is constrained by the Markov condition, and hence its capacity cannot exceed the Shannon capacity of the channel given by G. Since $\rho(G) = \rho(A)$ is the maximal spectral radius among all spectral radii of the strongly connected components of G, it suffices to show that we can find a Markov chain with $h(P, \mu) = h(G)$ in the case where G is strongly connected. We now show that we have equality for P of the form (6.4.5). Since G is strongly connected, it contains a cycle. Hence $\rho(A) \geq 1$. Since A is irreducible, the right and left eigenvectors $\mathbf{u}$ and $\mathbf{v}$ of A corresponding to $\rho(A)$ are unique positive vectors up to multiplication by a positive scalar. The assumption that $A\mathbf{u} = \rho(A)\mathbf{u}$ implies that P is stochastic and $G(P) = G$. Equalities (6.4.6) yield that μ is the unique stationary distribution corresponding to P. We now show that $h(P, \mu) = \log \rho(A)$. Indeed,

$$h(P,\mu) = -\sum_{i,j=1}^{n} \mu_i \frac{a_{ij}u_j}{\rho(A)u_i} \log \frac{a_{ij}u_j}{\rho(A)u_i}.$$

Expanding the logarithm, we get

$$h(P,\mu) = -\sum_{i,j=1}^{n} \mu_i \frac{a_{ij}u_j}{\rho(A)u_i} (\log a_{ij} + \log u_j - \log \rho(A) - \log u_i).$$

Since $a_{ij} \in \{0,1\}$, it follows that $a_{ij} \log a_{ij} = 0$, and hence the terms involving $a_{ij} \log a_{ij}$ vanish. We then have

$$-\sum_{i,j=1}^{n} \mu_i \frac{a_{ij}u_j}{\rho(A)u_i} \log u_j = -\sum_{i=1}^{n} \mu_i \left(\sum_{j=1}^{n} \frac{a_{ij}u_j}{\rho(A)} \log u_j \right).$$

Since $\sum_{j=1}^{n} \frac{a_{ij}u_j}{\rho(A)} = \frac{u_i}{\rho(A)}$,

$$-\sum_{i,j=1}^{n} \mu_i \frac{a_{ij}u_j}{\rho(A)u_i} \log u_i = -\sum_{i=1}^{n} \mu_i \log u_i.$$

Similarly,

$$-\sum_{i,j=1}^{n} \mu_i \frac{a_{ij}u_j}{\rho(A)u_i} \log u_j = -\sum_{j=1}^{n} v_j \log u_j.$$

Putting it all together, we get

$$h(P,\mu) = \log \rho(A). \qquad \square$$

To show that this is the only case where equality occurs is more complicated. This result follows from a more general theorem known as *Parry's theorem* [34].

ℹ 6.5 Pressure

View a word $\mathbf{a}_m = (a_1, \ldots, a_m) \in [n]^m$ as a molecule of length m arranged linearly on the lattice $\mathbb{N} \subset \mathbb{R}$, where the position i is occupied by an atom of type $a_i \in [n]$, $i = 1, \ldots, m$. Thus we have n different kinds of atoms. We refer to such a word $\mathbf{a}_m$ as an *m-configuration*. The set of all *unrestricted m-configurations* is $[n]^m$.

Assume that each atom of type i has potential e^{u_i} for some fixed $u_i \in \mathbb{R}$, $i = 1, \ldots, n$. The potential of an *m*-configuration $\mathbf{a}_m$ is then given by $\prod_{i=1}^{m} e^{u_{a_i}}$. Let $\mathbf{u} = (u_1, \ldots, u_n)^T \in \mathbb{R}^n$. Given a digraph $G = ([n], E)$ that contains a cycle, let $W(m)$ be the set of all G-permissible words of length m, i.e., all permissible walks on G of length m. Then the *grand partition function*, a concept from statistical mechanics, is the sum of all potentials of all G-permissible m-configurations:

$$Z(m, \mathbf{u}, G) := \sum_{\mathbf{a}_m = (a_1, \ldots, a_m) \in W(m)} \prod_{i=1}^{m} e^{u_{a_i}}. \tag{6.5.1}$$

A function $f : \mathbb{R}^n \to \mathbb{R}$ is called *nondecreasing* if for from $\mathbf{u} = (u_1, \ldots, u_n)^\top \leq \mathbf{v} = (v_1, \ldots, v_n)^\top$ (i. e., $u_i \leq v_i$ for $i = 1, \ldots, n$) it follows that $f(\mathbf{u}) \leq f(\mathbf{v})$. Clearly, $\prod_{i=1}^{m} e^{u_{a_i}}$ is a nondecreasing function on $\mathbb{R}^n$ for all $(a_1, \ldots, a_m) \in [n]^m$. Hence $Z(m, \mathbf{u}, G)$ and $\log Z(m, \mathbf{u}, G)$ are nondecreasing functions.

Let $C \subset \mathbb{R}^n$ be a convex set. A function $f : C \to \mathbb{R}$ is called a *convex* function if $f(t\mathbf{u} + (1 - t)\mathbf{v}) \leq tf(\mathbf{u}) + (1 - t)f(\mathbf{v})$ for all $\mathbf{u}, \mathbf{v} \in C$ and $t \in [0, 1]$. It is known that $\log Z(m, \mathbf{u}, G)$ is a convex function on $\mathbb{R}^n$ [27]. In the following theorem, we define a pressure function and investigate some of its properties. See [17] for a more detailed treatment.

Theorem 6.6. *Let* $G = ([n], E)$ *be a directed graph with a cycle. Then the sequence* $\log Z(m, \mathbf{u}, G)$, $m = 1, 2, \ldots$, *consists of nondecreasing convex functions, which are subadditive for each fixed* $\mathbf{u} \in \mathbb{R}^n$. *The pressure* $P(\mathbf{u}, G)$ *is defined as the limit*

$$P(\mathbf{u}, G) = \lim_{m \to \infty} \frac{1}{m} \log Z(m, \mathbf{u}, G),$$

which is a nondecreasing convex Lipschitz function on $\mathbb{R}^n$:

$$\left| P((u_1, \ldots, u_n), G) - P((v_1, \ldots, v_n), G) \right| \leq \max_{i \in [1, n]} |u_i - v_i|. \tag{6.5.2}$$

Denote by $A = (a_{ij})_{i,j=1}^n \in \{0, 1\}^{n \times n}$ *the adjacency matrix of* G. *Let*

$$A(\mathbf{u}) := \left(a_{ij} e^{\frac{u_i + u_j}{2}} \right)_{i,j=1}^n \in \mathbb{R}_+^{n \times n}.$$

Denote by $\rho(\mathbf{u})$ *the spectral radius of* $A(\mathbf{u})$. *Then* $P(\mathbf{u}, G) = \log \rho(\mathbf{u})$. *If* G *is strongly connected, then* $\log \rho(\mathbf{u})$ *is a smooth function of* $\mathbf{u}$.

Proof. The sequence $\log Z(m, \mathbf{u}, G), m = 1, \ldots$, is subadditive for the same reason $\log l_m = \log Z(m, \mathbf{0}, G)$ is subadditive. Hence the sequence $\frac{1}{m} \log Z(m, \mathbf{u}, G)$ converges to $P(\mathbf{u}, G)$ for each $\mathbf{u} \in \mathbb{R}^n$. Since each $\frac{1}{m} \log Z(m, \mathbf{u}, G)$ is nondecreasing and convex on $\mathbb{R}^n$, it follows that $P(\mathbf{u}, G)$ is nondecreasing and convex. Fix $\mathbf{u} = (u_1, \ldots, u_n)^\top$ and $\mathbf{v} = (v_1, \ldots, v_n)^\top \in \mathbb{R}^n$. Let $t = \max_{i \in [1, n]} |u_i - v_i|$. Then $v_i - t \leq u_i \leq v_i + t$ for all $i = 1, \ldots, n$. Hence

$$Z(m, \mathbf{v}, G)e^{-mt} \leq Z(m, \mathbf{u}, G) \leq Z(m, \mathbf{v}, G)e^{mt}$$

$$\Rightarrow \left| \frac{1}{m} \log Z(m, \mathbf{u}, G) - \frac{1}{m} \log Z(m, \mathbf{v}, G) \right| \leq t.$$

Let $m \to \infty$ and deduce (6.5.2). We now compare $\mathbf{1}^{\top} A(\mathbf{u})^{m-1}\mathbf{1}$ with $Z(m, \mathbf{u}, G)$. One term in $\mathbf{1}^{\top} A(\mathbf{u})^{m-1}\mathbf{1}$ is of the form

$$e^{\frac{u_{i_1}}{2}} a_{i_1 i_2} e^{\frac{u_{i_2}}{2}} e^{\frac{u_{i_2}}{2}} a_{i_2 i_3} e^{\frac{u_{i_3}}{2}} \dots e^{\frac{u_{i_{m-1}}}{2}} a_{i_{m-1} i_m} e^{\frac{u_{i_m}}{2}}.$$

If $(i_1, i_2, \dots, i_m) \notin W(m)$, then this product is equal to zero. If $(i_1, i_2, \dots, i_m) \in W(m)$, then this product is equal to $e^{-\left(\frac{u_{i_1}}{2} + \frac{u_{i_m}}{2}\right)} \prod_{j=1}^{m} e^{u_{i_j}}$. Let $t = \max_{i \in [1,n]} |u_i|$. Then $Z(m, \mathbf{u}, G)e^{-t} \leq \mathbf{1}^{\top} A^{m-1}\mathbf{1} \leq Z(m, \mathbf{u}, G)e^{t}$. Take the logarithm in all the terms of this inequality, let $m \to \infty$, and use (5.10.2) to deduce the equality $P(\mathbf{u}, G) = \log \rho(A(\mathbf{u}))$.

Assume finally that G is strongly connected. Then A is irreducible, and hence $A(\mathbf{u})$ is irreducible. So $\rho(\mathbf{u}) > 0$ is a simple root of the characteristic equation of $\det(zI - A(\mathbf{u}))$. Note that coefficients of this characteristic polynomial are analytic functions of $\mathbf{u}$. Hence the implicit function theorem (refer to Appendix A.7.3) yields that $\rho(\mathbf{u})$ and hence $\log \rho(\mathbf{u})$ are smooth functions of $\mathbf{u}$. □

For $m \in \mathbb{N}$, denote

$$C_n(m) := \{\mathbf{c} = (c_1, \dots, c_n) \in \mathbb{Z}_{+}^{n} : c_1 + \cdots + c_n = m\}.$$

For any G-permissible word $\mathbf{a}_m = (a_1, \dots, a_m) \in W(m)$, let $\mathbf{c}(\mathbf{a}_m) = (c_1, \dots, c_n) \in C_n(m)$, the frequency vector of the letter distributions in $\mathbf{a}_m$, that is, c_i is the number the letter i appears in $\mathbf{a}_m$. For any $\mathbf{c} \in C_n(m)$, set

$$W(\mathbf{m}, \mathbf{c}) := \{\mathbf{a}_m \in W(\mathbf{m}) : \mathbf{c}(\mathbf{a}_m) = \mathbf{c}\} \quad \text{for all } \mathbf{c} \in C_n(m).$$

This is the set of G-permissible words of $\mathbf{a}_m \in W(m)$ with color frequency vector $\mathbf{c}$.

Definition 6.7. $\mathbf{p} \in \Pi_n$ is called a density point of $[n]^{\mathbb{N}}(G)$ if there exist an increasing sequence of natural numbers $m_q \in \mathbb{N}$ and color frequency vectors $\mathbf{c}_q \in C_n(m_q)$ such that

$$m_q \to \infty, \quad W(m_q, \mathbf{c}_q) \neq \emptyset \quad \forall q \in \mathbb{N}, \quad \text{and} \quad \lim_{q \to \infty} \frac{\mathbf{c}_q}{m_q} = \mathbf{p}. \tag{6.5.3}$$

We denote by Π_G the set of all density points of $[n]^{\mathbb{N}}(G)$. For $\mathbf{p} \in \Pi_G$, we define

$$h_G^*(\mathbf{p}) := \sup_{m_q, \mathbf{c}_q} \limsup_{q \to \infty} \frac{\log |W(m_q, \mathbf{c}_q)|}{m_q} \geq 0, \tag{6.5.4}$$

where the supremum is taken over all sequences satisfying (6.5.3). We can think of $h_G^*(\mathbf{p})$ as the entropy for the color density $\mathbf{p}$.

It is not difficult to show (using a variant of the Cantor diagonal argument) that Π_G is a closed set (see Worked-out problem 6.7-1.) Furthermore, we may argue that h_G^* is *upper semicontinuous* on Π_G (see Problem 6.8-2.)

Theorem 6.8. *Let $G = ([n], E)$ be a strongly connected graph. Let $P(\cdot, G) : \mathbb{R}^n \to \mathbb{R}$ be the pressure function associated with G. Then*

$$\nabla P(\mathbf{u}) := \left(\frac{\partial P}{\partial u_1}(\mathbf{u}), \ldots, \frac{\partial P}{\partial u_n}(\mathbf{u}) \right) \in \Pi_n$$

is a distribution for each $\mathbf{u} \in \mathbb{R}^n$. Furthermore, for each $\mathbf{u} \in \mathbb{R}^n$, $\nabla P(\mathbf{u}) \in \Pi_G$, and

$$h_G^*(\nabla P(\mathbf{u})) = P(\mathbf{u}) - \nabla P(\mathbf{u}) \cdot \mathbf{u}, \tag{6.5.5}$$

$$P(\mathbf{u}) = \max_{\mathbf{p} \in \Pi_G}(\mathbf{p} \cdot \mathbf{u} + h_G^*(\mathbf{p})). \tag{6.5.6}$$

Proof. We first show that $\nabla P(\mathbf{u}) \in \Pi_n$. Since $P(\mathbf{u}, G)$ is nondecreasing, it follows that $\nabla P(\mathbf{u}) \geq \mathbf{0}$. Let $f(t, \mathbf{u}) = P(\mathbf{u} + t\mathbf{1})$ for $t \in \mathbb{R}$. From the definition of $A(\mathbf{u})$ it follows that $A(\mathbf{u} + t\mathbf{1}) = e^t A(\mathbf{u})$. Hence

$$f(t, \mathbf{u}) = P(\mathbf{u} + t\mathbf{1}) = \log \rho(A(\mathbf{u} + t\mathbf{1})) = t + \log \rho(A(\mathbf{u})) = t + P(\mathbf{u}).$$

Fix $\mathbf{u}$ and take the derivative with respect to t. The chain rule implies

$$1 = \frac{df(t, \mathbf{u})}{dt} = \nabla P(\mathbf{u}) \cdot \mathbf{1},$$

entailing $\nabla P(\mathbf{u}) \in \Pi_n$.

We now show the inequality

$$P(\mathbf{u}) \geq \mathbf{pu} + h_G^*(\mathbf{p}) \quad \text{for all } \mathbf{p} \in \Pi_G \text{ and } \mathbf{u} \in \mathbb{R}^n. \tag{6.5.7}$$

Fix $\mathbf{p} \in \Pi_G$ and let $m_q, \mathbf{c}_q, q \in \mathbb{N}$, be sequences satisfying (6.5.3). We have $Z(m_q, \mathbf{u}, G) \geq |W(m_q, \mathbf{c}_q)|e^{\mathbf{c}_q \mathbf{u}}$, since the right-hand side is just a partial sum of the sum represented by the left-hand side. Take logarithms, divide by m_q, take $\limsup_{q \to \infty}$, and use the definition of $P(\mathbf{u})$ and the limit in (6.5.3) to deduce $P(\mathbf{u}) \geq \mathbf{pu} + \limsup_{q \to \infty} \frac{\log |W(m_q, \mathbf{c}_q)|}{m_q}$. Now take the supremum over all sequences $m_q, \mathbf{c}_q$ satisfying (6.5.3) and use (6.5.4) to obtain (6.5.7). Hence

$$P(\mathbf{u}, G) \geq \sup_{\mathbf{p} \in \Pi_G} \mathbf{pu} + h_\Gamma^*(\mathbf{p}). \tag{6.5.8}$$

We now show that for each $\mathbf{u} \in \mathbb{R}^n$, there exists $\mathbf{p}(\mathbf{u}) \in \Pi_G$ such that

$$P(\mathbf{u}, G) = \mathbf{p}(\mathbf{u})\mathbf{u} + h_G^*(\mathbf{p}(\mathbf{u})). \tag{6.5.9}$$

Observe first that

$$|\mathcal{C}_n(m)| = \binom{m+n-1}{n-1} = O(m^{n-1}), \quad m \to \infty.$$

Then for each $m \in \mathbb{N}^d$,

$$Z(m, \mathbf{u}, G) = O(m^{n-1}) \max_{\mathbf{c} \in \Pi_n(m)} |W(m, \mathbf{c})| e^{\mathbf{c}^\top \mathbf{u}}.$$

Let

$$\mathbf{c}(m, \mathbf{u}) := \arg\max_{\mathbf{c} \in \Pi_n(m)} |W(m, \mathbf{c})| e^{\mathbf{cu}}. \tag{6.5.10}$$

Then

$$Z(m, \mathbf{u}, G) = O(m^{n-1}) |W(m, \mathbf{c}(m, \mathbf{u}))| e^{\mathbf{c}(m, \mathbf{u}) \mathbf{u}}. \tag{6.5.11}$$

Choose a sequence m_q such that $\frac{\mathbf{c}(m_q, \mathbf{u})}{m_q}$ converges to some $\mathbf{p}(\mathbf{u})$. By Definition 6.7 we have $\mathbf{p}(\mathbf{u}) \in \Pi_G$. Applying (6.5.11) to m_q and using the definitions of $P(\mathbf{u}, G)$ and $h_G^*(\mathbf{p}(\mathbf{u}))$, we can deduce

$$P(\mathbf{u}, G) \le \mathbf{p}(\mathbf{u}) \cdot \mathbf{u} + \limsup_{q \to \infty} \frac{\log |W(m_q, \mathbf{c}(m_q))|}{m_q} \le \mathbf{p}(\mathbf{u}) \cdot \mathbf{u} + h_G^*(\mathbf{p}(\mathbf{u})).$$

Combine this inequality with (6.5.8) to deduce (6.5.9) and (6.5.6).

It remains to show that $\nabla P(\mathbf{u}) = \mathbf{p}(\mathbf{u})$. Let $\mathbf{v} \in \mathbb{R}^n$ and $t \in \mathbb{R}$. Since $\mathbf{p}(\mathbf{u}) \in \Pi_G$, inequality (6.5.8), combined with equality (6.5.9), yields

$$P(\mathbf{u} + t\mathbf{v}, G) \ge \mathbf{p}(\mathbf{u}) \cdot (\mathbf{u} + t\mathbf{v}) + h_G^*(\mathbf{p}(\mathbf{u})) = t\mathbf{p}(\mathbf{u}) \cdot \mathbf{v} + P(\mathbf{u}, G).$$

Therefore

$$P(\mathbf{u} + t\mathbf{v}, G) - P(\mathbf{u}, G) \ge t\mathbf{p}(\mathbf{u}) \cdot \mathbf{v}.$$

Assume that $t > 0$. Divide by t and let $t \searrow 0$ to deduce that $\nabla P(\mathbf{u}) \cdot \mathbf{v} \ge \mathbf{p}(\mathbf{u}) \cdot \mathbf{v}$. Now assume that $t < 0$. Divide by t and let $t \nearrow 0$ to deduce that $\nabla P(\mathbf{u}) \cdot \mathbf{v} \le \mathbf{p}(\mathbf{u}) \cdot \mathbf{v}$. Hence $\nabla P(\mathbf{u}) \cdot \mathbf{v} = \mathbf{p}(\mathbf{u}) \cdot \mathbf{v}$ for all $\mathbf{v} \in \mathbb{R}^n$, which implies that $\nabla P(\mathbf{u}) = \mathbf{p}(\mathbf{u})$.

Ultimately, use (6.5.9) to deduce (6.5.5). $\qquad\square$

Proposition 6.9. *Let $G = ([n], E)$ be a strongly connected graph. For $\mathbf{u} \in \mathbb{R}^n$, let $A(\mathbf{u}) \in \mathbb{R}_+^{n \times n}$ be defined as in Theorem 6.6. Assume that $\mathbf{x}(\mathbf{u}) = (x_1(\mathbf{u}), \ldots, x_n(\mathbf{u}))^\top$ and $\mathbf{y}(\mathbf{u}) = (y_1(\mathbf{u}), \ldots, y_n(\mathbf{u}))^\top \in \mathbb{R}_+^n$ are positive left and right eigenvectors of $A(\mathbf{u})$, respectively, satisfying $A(\mathbf{u})\mathbf{x}(\mathbf{u}) = \rho(\mathbf{u})\mathbf{x}(\mathbf{u})$ and $A(\mathbf{u})^\top \mathbf{y}(\mathbf{u}) = \rho(\mathbf{u})\mathbf{y}(\mathbf{u})$, normalized by the condition $\mathbf{y}(\mathbf{u})^\top \mathbf{x}(\mathbf{u}) = 1$. Then $\nabla P(\mathbf{u}) = (y_1(\mathbf{u})x_1(\mathbf{u}), \ldots, y_n(\mathbf{u})x_n(\mathbf{u}))$ for all $\mathbf{u} \in \mathbb{R}^n$.*

Proof. Since $\rho(\mathbf{u}) > 0$ is a simple root of $\det(zI - A(\mathbf{u}))$, we can choose $\mathbf{x}(\mathbf{u})$ and $\mathbf{y}(\mathbf{u})$ to be analytic in $\mathbf{u}$ on $\mathbb{R}^n$. For example, first, choose $\mathbf{x}(\mathbf{u})$ and $\tilde{\mathbf{y}}(\mathbf{u}) \in \mathbb{R}_+^n$ to be the unique left and right eigenvectors of $A(\mathbf{u})$, respectively, each normalized to have unit length. Then let $\mathbf{y}(\mathbf{u}) = \frac{1}{\tilde{\mathbf{y}}(\mathbf{u})^\top \mathbf{x}(\mathbf{u})} \tilde{\mathbf{y}}(\mathbf{u})$.

Let ∂_i denote the partial derivative with respect to u_i. We have

$$\mathbf{y(u)}^\top \mathbf{x(u)} = 1 \quad \text{for all } \mathbf{u} \in \mathbb{R}^n \Rightarrow \partial_i\mathbf{y(u)}^\top\mathbf{x(u)} + \mathbf{y(u)}^\top\partial_i\mathbf{x(u)} = 0 \quad \text{for } i = 1,\dots,n.$$

Next, observe that $\mathbf{y(u)}^\top A(\mathbf{u})\mathbf{x(u)} = \rho(\mathbf{u})$. Taking the partial derivative with respect to u_i, we get

$$\partial_i\rho(\mathbf{u}) = \partial_i\mathbf{y(u)}^\top A(\mathbf{u})\mathbf{x(u)} + \mathbf{y(u)}^\top A(\mathbf{u})\partial_i\mathbf{x(u)} + \mathbf{y(u)}^\top\partial_i A(\mathbf{u})\mathbf{x(u)}$$
$$= \rho(\mathbf{u})(\partial_i\mathbf{y(u)}^\top\mathbf{x(u)} + \mathbf{y(u)}^\top\partial_i\mathbf{x(u)}) + \rho(\mathbf{u})y_i(\mathbf{u})x_i(\mathbf{u}).$$

Recalling the equality $P(\mathbf{u}, G) = \log\rho(\mathbf{u})$, we deduce that

$$\nabla P(\mathbf{u}) = (y_1(\mathbf{u})x_1(\mathbf{u}),\dots,y_n(\mathbf{u})x_n(\mathbf{u})).$$

Since $h(G) = P(\mathbf{0}, G)$, the characterization (6.5.6) yields

$$h(G) = \max_{\mathbf{p}\in\Pi_G} h_G^*(\mathbf{p}). \tag{6.5.12}$$

Let $A \in [0,1]^{n\times n}$ be an irreducible matrix. Let $\mathbf{x} = (x_1,\dots,x_n)^\top$ and $\mathbf{y} = (y_1,\dots,y_n)^\top \in \mathbb{R}_+^n$ be the right and left eigenvectors of A corresponding to $\rho(A)$, normalized by the condition $\mathbf{y}^\top\mathbf{x} = 1$. The distribution $(y_1x_1,\dots,y_nx_n) \in \Pi_n$ appears naturally in various contexts. For example, the Friedland–Karlin characterization of $\log\rho(A)$ is given by [18]

$$\log\rho(A) = \min_{\mathbf{z}=(z_1,\dots,z_n)^\top>0} \sum_{i=1}^n y_ix_i \log\frac{(A\mathbf{z})_i}{z_i}. \tag{6.5.13}$$

We now apply Theorem 6.8 to the model on $[2]^{\mathbb{N}}(G)$, where $A(G) = \left(\begin{smallmatrix}0&1\\1&1\end{smallmatrix}\right)$. (Here 0 is identified with 2.) Let $\mathbf{u} = (s,t)^\top$. Then $\nabla P(\mathbf{u}, G) = (p_1(\mathbf{u}), p_2(\mathbf{u})) \in \Pi_2$, and it follows that $p_2(\mathbf{u}) = 1 - p_1(\mathbf{u})$. We only need to consider $\mathbf{u} = (s,0)$, where $p_1(s) = \frac{dP((s,0)^\top, G)}{ds}$.
Thus $p := p_1(s)$ represents the density of 1s in all configurations of infinite strings of 0s and 1s, where no two 1s are adjacent. Clearly, $A(\mathbf{u}) = \left(\begin{smallmatrix}0&e^{\frac{s}{2}}\\e^{\frac{s}{2}}&1\end{smallmatrix}\right)$. Hence

$$\rho(\mathbf{u}) = \frac{1+\sqrt{1+4e^s}}{2},$$
$$p_1(s) = \frac{2e^s}{(1+\sqrt{1+4e^s})\sqrt{1+4e^s}} \frac{2}{(e^{-\frac{s}{2}}+\sqrt{e^{-s}+4})\sqrt{e^{-s}+4}}$$
$$= \frac{1}{2}\left(1-\frac{1}{\sqrt{1+4e^s}}\right) \in \left(0,\frac{1}{2}\right).$$

Note that p_1 is increasing on $\mathbb{R}$, with $p_1(-\infty) = 0$ and $p_1(\infty) = \frac{1}{2}$. As $P(\mathbf{0}) = h(G) = \log\frac{1+\sqrt{5}}{2}$, it follows that the value $p^* := p_1(0) = \frac{2}{(1+\sqrt{5})\sqrt{5}} \approx 0.2763932024$ is the density p^* of 1s for which $h(G) = h_G^*(p^*)$.

To derive the formula for $h_G(p)$, first note that if $p = p_1(s)$, then

$$\sqrt{1 + 4e^s} = \frac{1}{1 - 2p}, \quad s(p) = \log \frac{p(1-p)}{(1-2p)^2}.$$

Therefore

$$h_G^*(p) = \log \frac{1-p}{1-2p} - p \log \frac{p(1-p)}{(1-2p)^2}, \quad p \in \left(0, \frac{1}{2}\right). \qquad \square$$

6.6 Simulation of the gradient of pressure

In this section, we show that the gradient of a pressure is easy to simulate. Define

$$P_m(\mathbf{u}, G) := \frac{\log Z(m, \mathbf{u}, G)}{m}, \quad m \in \mathbb{N}. \tag{6.6.1}$$

Then

$$\frac{\partial P_m(\mathbf{u}, G)}{\partial u_i} = \frac{1}{m} \sum_{\phi \in W(m)} c_i(\phi) \frac{e^{\mathbf{c}(\phi)\mathbf{u}}}{Z(m, \mathbf{u}, G)}. \tag{6.6.2}$$

We introduce the following probability on $W(m)$, which depends on $\mathbf{u}$:

$$\mathbf{Pr_u}(\phi) := \frac{e^{\mathbf{c}(\phi)\mathbf{u}}}{Z(m, \mathbf{u}, G)} \quad \text{for all } \phi \in W(m). \tag{6.6.3}$$

Let $X_{i,m} : W(m) \to \mathbb{Z}_+$ be the random variable that counts the number of atoms (letters) of type i in the state $\phi \in W(m)$, that is, $X_{i,m} = c_i(\phi)$ for $i = 1, \ldots, n$. Let $\mathbf{X}_m := (X_{1,m}, \ldots, X_{n,m}) : W(m) \to \mathbb{Z}_+^m$ be a random vector variable. Then $\mathbf{X}_m(\phi) = \mathbf{c}(\phi)$. Using (6.6.1) and (6.6.3), we deduce that

$$\nabla P_m(\mathbf{u}, G) = \frac{1}{m} E_{\mathbf{u}}(\mathbf{X}_m), \quad m \in \mathbb{N}. \tag{6.6.4}$$

Note that in Problem 6.8-1, we consider computing the expected number of 1s in the configurations. This quantity corresponds to $E_0(X_{1,m})$. The second part of Problem 6.8-1, specifically part 1d, is to demonstrate that $\frac{E_0(X_{1,m})}{m}$ converges to a limit. It can be shown that this limit is given by

$$p_1(0) = \frac{2}{(1 + \sqrt{5})\sqrt{5}},$$

where $p_1(s)$ is defined in the last part of Section 6.5. More precisely, if G is strongly connected, it can be shown that

$$\lim_{m \to \infty} \nabla P_m(\mathbf{u}, G) = \nabla P(\mathbf{u}, G).$$

To approximate $\nabla P_m(\mathbf{u}, G)$, we simulate a random walk on G using a Markov chain as follows. (We assume that G is strongly connected and $m > 1$.)

Variant 1: Start the program with $w = e_1 = \cdots = e_n = 0$.

Let $P = (p_{\mathbf{a}_m \mathbf{b}_m})_{\mathbf{a}_m, \mathbf{b}_m \in \mathcal{W}(m)}$ be the stochastic matrix on the graph $G_m = (\mathcal{W}(m), E_m)$, where the vertices are indexed by all permissible m-walks $\mathcal{W}(m)$. The edge $(\mathbf{a}_m, \mathbf{b}_m) \in E_m$ if and only if $\mathbf{a}_m = (a_1, \ldots, a_m)$ and $\mathbf{b}_m = (b_1, \ldots, b_m)$ differ in at most one vertex k for some $k = 1, \ldots, n$.

Clearly, G_m is reversible, meaning that $(\mathbf{a}_m, \mathbf{b}_m) \in E_m$ if and only if $(\mathbf{b}_m, \mathbf{a}_m) \in E_m$. Assume that $a_1 = b_1, \ldots, a_{k-1} = b_{k-1}$ and $a_{k+1} = b_{k+1}, \ldots, a_m = b_m$.

Let $r = r(a_{k-1}, a_{k+1})$ be the number of all $j \in [n]$ such that

$$(a_{k-1}, j) \in E(G) \quad \text{and} \quad (j, a_{k+1}) \in E(G)$$

for all $j \in [n]$.

If $k = 1$, then the above condition reduces to $(j, a_2) \in E(G)$. If $k = m$, the condition reduces to $(a_{m-1}, j) \in E(G)$. For $a_k \neq b_k$ (and hence $a_i = b_i$ for $i \neq k$), we have

$$p_{\mathbf{a}_m \mathbf{b}_m} = \frac{1}{mr(a_{k-1}, a_{k+1})}.$$

The value of $p_{\mathbf{a}_m \mathbf{a}_m}$ is determined by the stochasticity condition. Therefore P is a symmetric stochastic matrix. If G_m is connected, then P has a unique uniform distribution. Although we do not generally know if G_m is connected, it is straightforward to show that G_m is connected for the model.

Our random walk on G_m is performed as follows:

1. First, generate a walk of length $\mathbf{a}_m$. Let $v = \mathbf{c}(\mathbf{a}_m) \cdot \mathbf{u} = c_1(\mathbf{a}_m)u_1 + \cdots + c_n(\mathbf{a}_m)u_n$, where $c_i(\mathbf{a}_m)$ is the number of i vertices in the walk given by $\mathbf{a}_m$. Compute $s = \exp(v)$, update $w = w + s$, and update $e_i = e_i + c_i(\mathbf{a}_m)s$ for $i = 1, \ldots, n$.
2. Move from $\mathbf{a}_m$ to $\mathbf{b}_m$ as follows: Choose a vertex a_k for some $k = 1, \ldots, m$ with probability $\frac{1}{m}$. Then choose $b_k = j$, where $(a_{k-1}, j) \in E(G)$ and $(j, a_{k+1}) \in E(G)$ with probability $\frac{1}{r(a_{k-1}, a_{k+1})}$. Thus $\mathbf{b}_m = (a_1, \ldots, a_{k-1}, b_k, a_{k+1}, \ldots, a_m)$.
3. Update $c_i(\mathbf{b}_m) = c_i(\mathbf{a}_m) - c_i(a_k) + c_i(b_k)$ for $i = 1, \ldots, n$, where $c_i(j)$ denotes the number i vertices in the color j (so $c_i(j) = \delta_{ij}$).
4. Recalculate s, w, and $\mathbf{e} = (e_1, \ldots, e_n)$ by replacing $\mathbf{a}_m$ with $\mathbf{b}_m$ in the above equalities and continue.

After completing the iterations, the vector $\frac{1}{w}\mathbf{e}$ provides an estimate of $E(\mathbf{X}_m)$, and $\frac{1}{mw}\mathbf{e}$ provides an estimate of $\nabla P_m(\mathbf{u}, G)$. If G_m is connected, then it is straightforward to show that these are valid estimates.

Variant 2: Start the program with $w = e_1 = \cdots = e_n = 0$.
Let $P = (p_{ij})_{i,j=1}^n$ with

$$p_{ij} = \frac{a_{ij}x_j}{\rho(G)x_i},$$

where $A(G) = (a_{ij})_{i,j=1}^n \in \{0,1\}^{n \times n}$ is the incidence matrix of G, and $\mathbf{x} = (x_1, \ldots, x_n)^\top \in \mathbb{R}_+^n$ is the positive eigenvector of $A(G)$ corresponding to the eigenvalue $\rho(G)$: $A(G)\mathbf{x} = \rho(G)\mathbf{x}$.

Choose a vertex a_1 at random with probability $\frac{1}{n}$. Then choose a neighbor of a_2 with probability $p_{a_1 a_2}$, and so on until a walk $\mathbf{a}_m = (a_1, \ldots, a_m) \in W(m)$ is generated. Let

$$v = \mathbf{c}(\mathbf{a}_m) \cdot \mathbf{u} = c_1(\mathbf{a}_m)u_1 + \cdots + c_n(\mathbf{a}_m)u_n,$$

where $c_i(\mathbf{a}_m)$ is the number of vertices of type i in the walk given by $\mathbf{a}_m$. Compute $s = \frac{x_{a_1} \exp(v)}{x_{a_m}}$, update $w = w + s$, and update $e_i = e_i + c_i(\mathbf{a}_m)s$ for $i = 1, \ldots, n$.

Pick a vertex a_{m+1} such that $(a_m, a_{m+1}) \in E(G)$, and define the new configuration $\mathbf{b}_m = (a_2, \ldots, a_{m+1})$. Then

$$c_i(\mathbf{b}_m) = c_i(\mathbf{a}_m) - c_i(a_1) + c_i(a_{m+1})$$

for $i = 1, \ldots, n$, where $c_i(j)$ denotes the number of vertices of type i in the color j (so $c_i(j) = \delta_{ij}$).

Recalculate s, w, and $\mathbf{e} = (e_1, \ldots, e_n)$ by replacing $\mathbf{a}_m$ with $\mathbf{b}_m$ in the above equalities and continue the process.

After completing the iterations, the vector $\frac{1}{w}\mathbf{e}$ provides an estimate of $E(\mathbf{X}_m)$, and $\frac{1}{mw}\mathbf{e}$ provides an estimate of $\nabla P_m(\mathbf{u}, G)$.

To explain why $\frac{1}{w}\mathbf{e}$ is an estimate of $E(\mathbf{X}_m)$ in this case, note that the probability of generating a walk $\mathbf{a}_m = (a_1, \ldots, a_m)$ is

$$\frac{1}{n}p_{a_1 a_2}p_{a_2 a_3} \cdots p_{a_{m-1}a_m} = \frac{1}{n}\frac{x_{a_m}}{\rho(G)^{m-1}x_{a_1}},$$

which explains why we use the product

$$s = \frac{x_{a_1}\exp(v)}{x_{a_m}}.$$

Thus $\frac{1}{w}\mathbf{e}$ is an estimate of $E(\mathbf{X}_m)$.

It is straightforward to show that, under the assumption that G is strongly connected, you can go from any $\mathbf{a}_m \in W(m)$ to any other $\mathbf{b}_m \in W(m)$ using the Markov chain described above in a finite number of steps. The disadvantage of Variant 2 is that it requires knowledge of the eigenvector $\mathbf{x}$ and eigenvalue $\rho(G)$.

Remark 6.10. Let $G = ([n], E)$ be a strongly connected graph, and let $A(G) = (a_{ij})_{i,j=1}^{n} \in \{0, 1\}^{n \times n}$ be the inci- $\boxed{!}$
dence matrix of G. Let $\mathbf{x} = (x_1, \ldots, x_n)^\top$ and $\mathbf{y} = (y_1, \ldots, y_n)^\top \in \mathbb{R}_+^n$ be the positive left and right eigenvectors
of $A(G)$ corresponding to $\rho(G)$: $A(G)\mathbf{x} = \rho(G)\mathbf{x}$ and $\mathbf{y}^\top A(G) = \rho(G)\mathbf{y}^\top$, normalized by the condition $\mathbf{y}^\top \mathbf{x} = 1$.
   Introduce the following probability measure on $[n]^m$, the set of walks of length m on the complete
digraph with n vertices:

$$\mathbf{Pr}\big(\mathbf{b}_m = (b_1, \ldots, b_m)\big) := \frac{1}{\rho(G)^{m-1}} y_{b_1} a_{b_1 b_2} \cdots a_{b_{m-1} b_m} x_{b_m} \quad \text{for all } b_1, \ldots, b_m \in [n]. \qquad (6.6.5)$$

   Then $\mathbf{Pr}(\mathbf{b}) = 0$ if $\mathbf{b} \notin W(m)$, and $\mathbf{Pr}(\mathbf{b}) = \frac{1}{\rho(G)^{m-1}} y_{b_1} x_{b_m}$ if $\mathbf{b}_m = (b_1, \ldots, b_m) \in W(m)$. This measure is
called the *Parry* measure and is an example of a *Gibbs* measure [30].

6.7 Worked-out problems $\boxed{?}$

1.   Let $G = ([n], E)$ be a digraph that contains a loop. Denote

$$[n]^{\mathbb{N}}(G) := \{\mathbf{a} = (a_1, \ldots) \in [n]^{\mathbb{N}} : (a_i, a_{i+1}) \in E, i = 1, 2, \ldots\}.$$

Denote by Π_G the set of all density points of $[n]^{\mathbb{N}}(G)$. Prove that Π_G is a closed set.

Solution. By Worked-out problem 4.3-1 we know that Π_n is a compact set. Clearly,
Π_G is contained in Π_n. Let $\mathbf{p}_k \in \Pi_G$, $k \in \mathbb{N}$, be a sequence that converges to $\mathbf{p}$.
By taking a subsequence we can assume that $|\mathbf{p} - \mathbf{p}_k| \le 1/k$ for $k \in \mathbb{N}$. Each $\mathbf{p}_k$
is a limit of the sequence $\mathbf{c}_{1,k}/m_{1,k}, \mathbf{bc}_{2,k}/m_{2,k}, \ldots$ for $k = 1, 2, \ldots$ Again, by taking a
subsequence, if needed, we can assume that $|\mathbf{p}_k - \mathbf{c}_{q,k}/m_{q,k}| < 1/q$ for $q \in \mathbb{N}$. The
diagonal sequence is $\mathbf{c}_{1,1}/m_{1,1}, \mathbf{c}_{2,2}/m_{2,2}, \ldots, \mathbf{c}_{q,q}, \ldots$. So $|\mathbf{p}_k - \mathbf{c}k, k/m_{k,k}| \le 1/k$. Hence
the diagonal sequence converges to $\mathbf{p}$. Thus Π_G is closed.

6.8 Problems $\boxed{i}$

1.   Consider the hard-core model on the chain $C_m := 1\text{—}2\text{—}\cdots\text{—}m$ as explained in
Section 6.1. Let l_m be the number of hard-core configurations on C_m, and let s_m denote the total number of 1s in these configurations. Denote the space of hard-core
configurations on C_m by Φ_m. We assume that $\mathbf{Pr}(\phi) = \frac{1}{l_m}$ for each $\phi \in \Phi_m$. Let
$X_m : \Phi_m \to \mathbb{Z}^+$ be the random variable that counts the number of 1s in the configuration $\phi \in \Phi_m$.
   (a) Prove the formula

$$s_m = s_{m-1} + s_{m-2} + l_{m-2}, \quad m = 3, \ldots. \qquad (6.8.1)$$

   (b) Show that

$$s_m = (mb_1 + b_2)\left(\frac{1 + \sqrt{5}}{2}\right)^m + (mb_3 + b_4)\left(\frac{1 - \sqrt{5}}{2}\right)^m, \quad m = 1, \ldots, \qquad (6.8.2)$$

satisfies the recurrence relation for s_m given in (6.8.1) for any l_m satisfying $l_m = l_{m-1} + l_{m-2}$, $m = 3,\dots$.

(c) Find the values of b_1, b_2, b_3, and b_4 appearing in (6.8.2) using the given values $s_1 = 1$, $s_2 = 2$, $s_3 = 5$, and $s_4 = 10$. You may use software such as MATLAB, Maple, or any other computational tool. Provide the values of b_1, b_2, b_3, and b_4 with a precision of four digits.

(d) Prove the following formulas:

$$\lim_{m \to \infty} \frac{\log l_m}{m} = \log\left(\frac{1 + \sqrt{5}}{2}\right),$$ (6.8.3)

and

$$\lim_{m \to \infty} \frac{E(X_m)}{m} = \frac{10b_1}{5 + 3\sqrt{5}}.$$ (6.8.4)

Find the values appearing in these formulas with a precision of four digits.

(e) Write a computer program to:
  i. For $m = 2,\dots,10$, enumerate all the hard-core configurations. Verify the results using the recursive formula for l_m. For each m, print l_m and $\frac{1}{l_m}$.
  ii. Sum the number of 1s in all the configurations and check if it matches s_m computed from the recursive formula. For each m, print $E(X_m) = \frac{s_m}{l_m}$.

(f) Implement the **MCMC** algorithm as explained in Example 6.2 for the chain C_m. (Note that Example 6.2 deals with a square grid, so you need to apply it to the one-dimensional case.) Run this **MCMC** for $m = 2,\dots,10$ for approximately $100 \times l_m$ iterations for each m and perform the following:
  i. Track the total number N of configurations visited. In every configuration visited, count the number of 1s and keep a running total M. Print the ratio M/N, an estimate of $E(X_m)$.
  ii. Keep track of the states (hard-core configurations) visited. For each configuration $\phi \in C_m$, compute the number $L(\phi)$ of times ϕ was visited. Calculate the ratio $L(\phi)/N$, an estimate for the stationary distribution for each ϕ. Let $K \subset l_m$ be the number of different states visited. (Do not count states that were never visited.) Assume that each visited state has a probability of $1/K$. Find the mean and standard deviation of the estimates $L(\phi)/N$ for the states visited. Find the mean, standard deviation, and the number of states never visited.

2. Let $f : A \to \mathbb{R}$, and let $a \in A$. It is said that f is *lower semicontinuous* at a if for every $\epsilon > 0$, there exists $\delta > 0$ such that

$$f(a) - \epsilon < f(x) \quad \text{for all } x \in B(a, \delta) \cap A.$$ (3.7.1)

Similarly, it is said that f is *upper semicontinuous* at a if for every $\epsilon > 0$, there exists $\delta > 0$ such that

$$f(x) < f(a) + \epsilon \quad \text{for all } x \in B(a, \delta) \cap A.$$

It is said that f is lower (upper) semicontinuous at A if for every $a \in A$, f is lower (upper) semicontinuous at a.

It is clear that f is continuous at a if and only if f is lower semicontinuous and upper semicontinuous at this point. Let $f : A \to \mathbb{R}$, and let $a \in A$ be a limit point of A.

(a) Prove that f is lower semicontinuous at a if and only if

$$\liminf_{x \to a} f(x) \geq f(a).$$

(b) Similarly, prove that f is upper semicontinuous at a if and only if

$$\limsup_{x \to a} f(x) \leq f(a).$$

(c) Let $h_G^*(\mathbf{p})$ be as in Definition 6.7. Prove that h_G^* is upper semicontinuous on Π_G.

7 Hausdorff dimension of subshifts of finite type

ℹ ## 7.1 Pseudometrics on infinite graph sequences

In this section, we introduce the essential concepts and terminologies, assuming a basic understanding of topology and linear algebra. For additional background, the reader is directed to Appendices A.7, A.9, and A.10. For a detailed discussion on the ergodic theory of continuous transformations on compact metrizable spaces, we refer the reader to [44]. See also [13, 14] for further insights into the Hausdorff dimension of subshifts and discrete Lyapunov exponents. We begin by considering a setup similar to that discussed in Section 6.1.

Let $[n] = \{1, \dots, n\}$ represent an alphabet with n letters. Consider a directed graph $G = ([n], E)$, where $E \subset [n] \times [n]$. Denote by $A(G)$ the adjacency matrix of G and by $\rho(G) = \rho(A(G))$ its spectral radius. Define the following sets:

$$G^p = \{x = (x_i)_{i=1}^p : (x_i, x_{i+1}) \in E, \ i = 1, \dots, p-1\}, \quad p = 2, \dots;$$
$$G_1 = [n];$$
$$G^\infty = \{x = (x_i)_{i=1}^\infty : (x_i, x_{i+1}) \in E, \ i = 1, \dots\}.$$

Assume that $A(G)$ is not a nilpotent matrix. We consider G^∞ as a compact topological space under the standard Tychonoff (product) topology. A function $d : G^\infty \times G^\infty \to \mathbb{R}^+$ is called a *pseudometric* if there exists a metric $\tilde{d}$ on G^∞ such that

$$\frac{d(a,b)}{K} \leq \tilde{d}(a,b) \leq Kd(a,b), \quad a, b \in G^\infty,$$

for some $K \geq 1$. Let $d, \tilde{d} : G^\infty \times G^\infty \to \mathbb{R}^+$. The functions d and $\tilde{d}$ are called *equivalent functions* if the above inequality holds. The set G^∞ is said to be *compact with respect to the pseudometric* d if G is compact with respect to an equivalent metric $\tilde{d}$. Let $B = (b_{ij})_{i,j=1}^n$ be an $n \times n$ positive matrix. Define the distance d_B satisfying:

- $d_B(a,a) = 0, a \in G^\infty$;
- $d_B(a,b) = d_B(b,a) = 1$ if $a = (a_i)_{i=1}^\infty, b = (b_i)_{i=1}^\infty \in G^\infty$, and $\max\{|a_1 - b_1|, |a_2 - b_2|\} > 0$;
- $d_B(a,b) = \prod_{i=1}^{p-2} b_{a_i a_{i+1}}$ if $a = (a_i)_{i=1}^\infty, b = (b_i)_{i=1}^\infty \in G^\infty, a_1 = b_1, \dots, a_{p-1} = b_{p-1}$, and $a_p \neq b_p, p \geq 3$.

The matrix B is said to have the *G-cycle property* if for any (directed) cycle in G,

$$\sigma = (i_j)_{j=1}^{k+1} \in G^{k+1} \quad \text{with } i_{k+1} = i_1 \text{ and } i_j \neq i_l \text{ for } 1 \leq j < l \leq k,$$

we have the inequality $\prod_{j=1}^k b_{i_j i_{j+1}} < 1$.

Assume that B is a positive matrix with the G-cycle property. We show that d_B is a pseudometric. Moreover, G^∞ is compact with respect to d_B. (If G is strongly connected

https://doi.org/10.1515/9783111337388-007

and G^∞ is an infinite set, then the assumption that d_B is a pseudometric such that G^∞ is compact with respect to d_B implies that B has the G-cycle property.)

Outline. In this chapter, we determine the exact value of the Hausdorff dimension, denoted by $\delta(G, B)$, for the set G^∞ with respect to a specific pseudometric. This dimension is equivalent to the *Hausdorff dimension* with respect to another metric $\tilde{d}$, given that the matrix elements b_{ij} are of the form $b_{ij} = \tau^{m_{ij}}$, where $m_{ij} \in \mathbb{Z}$, $(i,j) \in E = E(G)$, and $0 < \tau < 1$.

We then show that

$$\delta(G, B) = \frac{\log \rho(G'')}{\log \tau^{-1}},$$

where G'' is the corresponding digraph induced by G and is a subgraph of $[N] \times [N]$.

In cases where $m_{ij} \geq 1$ for $(i,j) \in E(G)$, the digraph G'' is constructed by adding $m_{ij} - 1$ extra vertices along any edge $(i,j) \in E(G)$, creating a directed path in G'' of length m_{ij}.

We apply these results to find the Hausdorff dimension of the limit set of a finitely generated free group of isometries acting on a locally finite tree. The reader is referred to [30] for a detailed discussion.

Historical note. The theory of Hausdorff measure and dimension was developed to address limitations in existing measures, such as Lebesgue measure, by defining a more general notion of size. The idea is to evaluate the size of a set using an α-dependent measure μ that targets sets of dimension α. Sets with a dimension less than α are considered "small" (having measure zero), whereas those with a dimension greater than α are considered "large" (having infinite measure). Although the Lebesgue measure achieves this for sets in $\mathbb{R}^d$, it only assigns integer values to dimensions, thus missing finer structural details. The Hausdorff measure generalizes this concept by considering the volume of coverings with rectangles and extending it to arbitrary metric spaces and fractional α.

Introduced by Felix Hausdorff (Figure 7.1) in 1918, the Hausdorff dimension is a measure of roughness or fractal dimension. For instance, the Hausdorff dimension is zero for a single point, 1 for a line segment, 2 for a square, and 3 for a cube. For shapes that are smooth or have few corners, those of traditional geometry, the Hausdorff dimension corresponds to the usual integer-valued topological dimension.

7.2 Pseudometrics on an SFT

Strict subshifts of finite type. Let G and its powers be as in Section 7.1. Set $G(1) := G$, and for $m > 1$, define $G(m) \subset G^m \times G^m$ as follows:

$$G(m) = \{(a, b) : a = (a_i)_{i=1}^m, b = (b_i)_{i=1}^m \in G^m \text{ and } a_{i+1} = b_i \text{ for } i = 1, \ldots, m-1\}.$$

Clearly, any $a \in G^m$ induces a unique element $a' \in G(m)^\infty$ and vice versa. Define the correspondence $\phi_m : G^m \to G(m)^\infty$ by $a \mapsto a'$.

Figure 7.1: Felix Hausdorff (1868–1942). Photo credit: Wikipedia. Source: https://en.wikipedia.org/wiki/Felix_Hausdorff.

A subset $X \subset G^m$ is called a *strict subshift of finite type*, abbreviated as **SSFT** if there exist $m \geq 1$ and a strict subset $\Delta \subset G(m)$, viewed as a digraph on G^m vertices, such that $\phi_m(X) = \Delta^\infty$.

Lemma 7.1. *Let* $d : G^\infty \times G^\infty \to \mathbb{R}^+$ *be a metric on* G^∞ *such that G is compact with respect to d. Suppose that d satisfies the B property for a positive matrix B:*

$$d((a_i)_{i=1}^\infty, (b_i)_{i=1}^\infty) = b_{pq}d((a_i)_{i=2}^\infty, (b_i)_{i=2}^\infty),$$

where $(a_i)_{i=1}^\infty, (b_i)_{i=1}^\infty \in G^\infty$, $a_1 = b_1 = p$, *and* $a_2 = b_2 = q$.
- *Then d_B is a pseudometric equivalent to d.*
- *Suppose furthermore that G is strongly connected and $\rho(G) > 1$. (That is, G does not consist of one cycle, and thus G^∞ is an infinite set.) Then B satisfies the G-cycle property.*

Proof. Suppose first that $a = (a_i)_{i=1}^\infty, b = (b_i)_{i=1}^\infty \in G^\infty$ with $a_1 \neq b_1$. Let D be the diameter of G^∞ in the metric d. Clearly, $d(a,b) \leq D \cdot d_B(a,b)$. The compactness argument shows that there exists $\epsilon_1 > 0$ such that $d(a,b) \geq \epsilon_1$ for any pair $a, b \in G^\infty$ that differs in the first coordinate. Hence $d(a,b) \geq \epsilon_1 \cdot d_B(a,b)$ for all pairs a, b that differ in the first coordinate. A similar argument shows that for some $\epsilon_2 > 0$,

$$\epsilon_2 \cdot d_B(a,b) \leq d(a,b) \leq D \cdot d_B(a,b),$$

where $a = (a_i)_{i=1}^\infty, b = (b_i)_{i=1}^\infty \in G^\infty$ with $a_2 \neq b_2$.
Set $K := \max(D, \epsilon_1^{-1}, \epsilon_2^{-1})$. We claim that

$$\frac{d(a,b)}{K} \leq d_B(a,b) \leq K \cdot d(a,b)$$

for d and d_B.

If $a \neq b \in G^\infty$ and a and b differ in the first or second coordinates, then the inequality follows directly from the previous arguments and the definition of K. For cases where a and b have exactly $p \geq 2$ common coordinates in the first positions, the B property of d and d_B can be used to deduce the desired inequality from the previous cases. Thus d_B is a pseudometric.

Assume that G is strongly connected and $\rho(G) > 1$. Let d be a metric that satisfies the conditions of the lemma, and suppose G^∞ is compact with respect to d. We claim that B satisfies the G-cycle property.

Consider a cycle $\sigma = (i_j)_{j=1}^{k+1}$, where $i_{k+1} = i_1$ and $1 \leq k \leq n$. Let $\ell(\sigma)$ be the length of the cycle σ, which is equal to k. Denote by Σ_n the set of all cycles in the complete digraph on n vertices: $([n], [n] \times [n])$. For an $n \times n$ nonnegative matrix B, define

$$B(\sigma) = \prod_{j=1}^{k} b_{i_j i_{j+1}},$$

$$\eta(B) = \max_{\sigma \in \Sigma_n} B^{\frac{1}{\ell(\sigma)}}(\sigma).$$

Note that B satisfies the G-cycle property if and only if $\eta(B) < 1$.

Since G is strongly connected and $\rho(G) > 1$, the digraph G contains an additional edge. By adjusting the starting vertex we can assume that $(i_1, l) \in G$ with $l \neq i_2$. Let $a = (a_j)_{j=1}^\infty \in G^\infty$ be a *periodic vector* with $a_j = i_l$ for $j \equiv l \pmod{k}$, that is, $a = (a_1, a_2, \ldots, a_k, a_1, a_2, \ldots, a_k, \ldots)$ with $a_{p+kl} = a_p$, for all $p \in \mathbb{N}$ and $l \in \mathbb{N}$. Assume that $(a_1, \ldots, a_k, a_1)$ is a cycle σ. Our assumptions imply the existence of the vector

$$b(q) = (b_j(q))_{j=1}^\infty \in G^\infty,$$

where $b_j(q) = a_j$ for $j = 1, \ldots, qk + 1$, and $b_{qk+2}(q) = l$. Clearly, the sequence $b(q)$, $q = 1, 2, \ldots$, converges to a in the Tychonoff topology. Since G^∞ is compact with respect to the metric d, it follows that $\lim_{q \to \infty} d(b(q), a) = 0$. Given that d_B is an equivalent pseudometric, it also follows that $\lim_{q \to \infty} d_B(b(q), a) = 0$. Next, observe that

$$d_B(b(q), a) = B(\sigma)^q.$$

Hence $B(\sigma) < 1$ for each cycle σ, which means that B satisfies the G-cycle property. □

We now show that if a nonnegative matrix B satisfies the G-cycle property, it can be replaced by a matrix $C = (c_{ij})_{i,j=1}^n$, where $0 \leq c_{ij} < 1$ are such that d_B and d_C are equivalent. We will need the following results from [11, Theorem 7.2 and Remark 7.3].

For a square $n \times n$ matrix $A = (a_{ij})_{i,j=1}^n$, denote $\|A\|_{\max} = \max_{i,j \in [n]} |a_{ij}|$.

Theorem 7.2. *Let $B = (b_{ij})_{i,j=1}^n$ be a nonnegative matrix. Denote by Δ_n the set of $n \times n$ diagonal matrices with positive diagonal entries. Then*

$$\inf_{D \in \Delta_n} \left\| D^{-1}BD \right\|_{\max} = \eta(B).$$

In particular, B satisfies the G(B)-cycle property if and only if there exists a nonnegative matrix $C = (c_{ij})_{i,j=1}^{n}$ with $c_{ij} < 1$ for $i,j = 1,\ldots,n$ that is diagonally similar to B.

Lemma 7.3. Let B and C be two $n \times n$ nonnegative nonnilpotent matrices that are diagonally similar. Suppose $G \subseteq G(B) = G(C)$ is a graph with $\rho(G) \geq 1$. Then d_B and d_C are equivalent. Additionally, if B satisfies the G-cycle property, then there exists a matrix $A = (a_{ij})_{i,j=1}^{n}$ with $0 \leq a_{ij} < 1$ that satisfies the following: $G(A) = G$ and $A(\sigma) = B(\sigma)$ for every cycle σ in G. In particular, d_A is a metric, d_B is a pseudometric on G^∞, and G^∞ is compact with respect to d_B. Finally, if there exists $0 < \tau < 1$ such that $b_{ij} = \tau^{m_{ij}}$ for $m_{ij} \in \mathbb{Z}$ and $(i,j) \in G$, then there can be constructed A such that

$$G(A) = G, \quad a_{ij} = t^{n_{ij}}, \quad \text{where } 1 \leq n_{ij} \in \mathbb{Z},$$

$$(i,j) \in G, \quad \text{and} \quad 0 < t = \tau^{\frac{1}{L}} < 1 \quad \text{for some } L \geq 1. \tag{7.2.1}$$

Proof. Assume initially that B and C are diagonally similar. Define the following constants:

$$K_1(B) = \min \left\{ \min_{2 \leq k \leq n,\, (a_i)_{i=1}^{k} \in G^k} \prod_{i=1}^{k-1} b_{a_i a_{i+1}}, 1 \right\},$$

$$K_2(B) = \max \left\{ \max_{2 \leq k \leq n,\, (a_i)_{i=1}^{k} \in G^k} \prod_{i=1}^{k-1} b_{a_i a_{i+1}}, 1 \right\},$$

$$K_1(C) = \min \left\{ \min_{2 \leq k \leq n,\, (a_i)_{i=1}^{k} \in G^k} \prod_{i=1}^{k-1} c_{a_i a_{i+1}}, 1 \right\},$$

$$K_2(C) = \max \left\{ \max_{2 \leq k \leq n,\, (a_i)_{i=1}^{k} \in G^k} \prod_{i=1}^{k-1} c_{a_i a_{i+1}}, 1 \right\}.$$

Let $a^q = (a_i)_{i=1}^{q} \in G^q$, where $q > 1$. Then a^q can be considered as a path of length $q-1$ in the digraph G. Any such path can be decomposed into cycles and a chain of length at most $n-1$. Since $B(\sigma) = C(\sigma)$, for all cycles σ in G, we have

$$\prod_{i=1}^{q-1} b_{a_i a_{i+1}} \leq \frac{K_2(B)}{K_1(C)} \prod_{i=1}^{q-1} c_{a_i a_{i+1}},$$

$$\prod_{i=1}^{q-1} c_{a_i a_{i+1}} \leq \frac{K_2(C)}{K_1(B)} \prod_{i=1}^{q-1} b_{a_i a_{i+1}}.$$

Thus

$$\frac{K_1(B)}{K_2(C)} d_C(a,b) \leq d_B(a,b) \leq \frac{K_2(B)}{K_1(C)} d_C(a,b) \quad \text{for } a,b \in G^\infty.$$

This proves the equivalence of d_B and d_C.

Now suppose that $G \subseteq G(B)$. Define $\hat{B} = (\hat{b}_{ij})^n_{i,j=1}$ such that

$$\hat{b}_{ij} = \begin{cases} 0 & \text{if } (i,j) \notin G, \\ b_{ij} & \text{if } (i,j) \in G. \end{cases}$$

Then $d_{\hat{B}} = d_B$. Assume that B satisfies the G-cycle property. Clearly, $G(\hat{B}) = G$, and $\hat{B}$ satisfies the G-cycle property. By applying Theorem 7.2 we can deduce the existence of a diagonal matrix $D = \text{diag}(d_1, \ldots, d_n)$ with $d_i > 0$ for $i = 1, \ldots, n$ such that $A = D\hat{B}D^{-1}$ and all the entries of A are less than 1. Invoking Problem 7.6-2, $A(\sigma) = B(\sigma)$ for any cycle σ in G. Furthermore, d_B is equivalent to d_A. Since d_A is a metric on G^∞, it follows that d_B is a pseudometric. Given that G^∞ is compact with respect to d_A, it follows that G^∞ is also compact with respect to d_B. (See Problem 7.6-1.) Assume further that the nonzero entries of $\hat{B}$ are integral powers of τ. We approximate each d_i with τ^{q_i}, where $q_i \in \mathbb{Q}$, so that the following condition holds: let $D_1 = \text{diag}(\tau^{q_1}, \ldots, \tau^{q_n})$ be such that all the entries of $A_1 = D_1 \hat{B} D_1^{-1}$ are less than 1. Then A_1 takes the form of (7.2.1). □

7.3 Hausdorff dimension of G^∞

We overview the notion of the Hausdorff dimension for the compact space G^∞ with a metric $\tilde{d}$ that is equivalent to pseudometric $d = d_B$, where B has the G-cycle property.

Let

$$D(a, \epsilon) = \{b \in G^\infty \mid \tilde{d}(a, b) < \epsilon\}.$$

Let $\mathcal{A} \subset G^\infty$ be a finite set of cardinality $|\mathcal{A}| = N$: $\mathcal{A} = \{a^1, \ldots, a^N\}$. An ϵ-$\mathcal{A}$ cover of G^∞ satisfies

$$G^\infty = \bigcup_{a^j \in \mathcal{A}} D(a^j, \epsilon_j), \quad 0 < \epsilon_j \leq \epsilon.$$

Denote

$$H^r_{\tilde{d}}(\epsilon) = \inf_{\epsilon\text{-}\mathcal{A}\text{cover}} \sum_{j \in [|\mathcal{A}|]} \epsilon_j^d.$$

Clearly, $H^r_{\tilde{d}}(\epsilon)$ is a decreasing function of ϵ. For $\epsilon \in [0,1]$, it is a decreasing function of r. Set

$$\lim_{\epsilon \searrow 0} H^r_{\tilde{d}}(\epsilon) = H^r_{\tilde{d}} \in [0, \infty],$$

$$\delta(\tilde{d}) = \inf\{r \mid H^r_{\tilde{d}} = 0\}.$$

Then $\delta(\tilde{d})$ is the Hausdorff dimension of G^∞.

It is straightforward to show that if in the above definitions, we replace $\tilde{d}$ by d, then we will have the equality $\delta(\tilde{d}) = \delta(d)$. Set

$$\delta(d) = \delta(d_B) \overset{\text{def}}{=} \delta(G, B).$$

Consider the matrix $J = (1)_{n\times n}$ with all entries equal to 1, and let $0 < \tau < 1$. The metric $d_{\tau J}$ is equivalent to the following standard metric:

$$d_\tau((a_i)_{i=1}^\infty, (b_i)_{i=1}^\infty) = \max_{1\leq i} \tau^{i-1}\mathrm{ham}(a_i, b_i),$$

where $(a_i)_{i=1}^\infty$ and $(b_i)_{i=1}^\infty$ are elements of G^∞. Notice that here ham $: \mathbb{Z} \times \mathbb{Z} \to \mathbb{Z}^+$ represents the Hamming distance, i. e., $\mathrm{ham}(x,y) = 1$ for $x \neq y$. The following result will be used to establish our formula for $\delta(G, B)$.

Lemma 7.4. *Let $G \subset [n] \times [n]$ and assume that $\rho(G) > 0$ if and only if $G^\infty \neq \emptyset$. Suppose $0 < \tau < 1$. Then*

$$\delta(G, \tau J) = \frac{\log \rho(G)}{\log \tau^{-1}}.$$

Proof. Assume that $\tau^{k+1} < \epsilon \leq \tau^k$ and suppose k is sufficiently large. Let $(a_i)_{i=1}^\infty, (b_i)_{i=1}^\infty \in G^\infty$ with $a_i = b_i$ for $i = 1, \ldots, p-1$ and $a_p \neq b_p$ for $2 \leq p$.

Then $d_{\tau J}(a, b) = \tau^{p-2}$. If $p < k + 3$, then it follows that $d_{\tau J}(a, b) \geq \epsilon$. Thus, for any $a \in G^\infty$, there exists $a^j \in A$ such that a and a^j share the same $k + 2$ coordinates. Hence the set $\mathcal{A}$ contains a subset $\mathcal{B}_{k+2}$ with the following property: for any $a \in G^\infty$, there is exactly one point $a^j \in \mathcal{B}_{k+2}$ that shares the same first $k + 2$ coordinates with a. On the other hand, any set $\mathcal{B}_{k+2}$ with this property will generate an ϵ-cover:

$$\bigcup_{a^j \in \mathcal{B}_{k+2}} D(a^j, \epsilon_j) = G^\infty, \quad \tau^{k+1} < \epsilon_j \leq \tau^k.$$

Let v_{k+2} be the cardinality of the set $\mathcal{B}_{k+2}$. Define

$$\Lambda_\epsilon^a = \inf\left\{ \sum_{a^j \in \mathcal{A}} \epsilon_j^a \mid \bigcup_{a^j \in \mathcal{A}} D(a^j, \epsilon_j) = G^\infty, \epsilon_j \leq \epsilon \right\}.$$

It follows that

$$\Lambda_\epsilon^a \leq v_{k+2}\epsilon^a \leq v_{k+2}\tau^{ka}, \quad \tau^{k+1} < \epsilon \leq \tau^k.$$

Note that v_p is the sum of all the entries of the matrix A_p. The asymptotic behavior of v_p is discussed in [19], where it is proved that it is precisely of order $Tp^l\rho(G)^p$ for $p \gg 1$, where $0 \leq l < n$ is an integer, and $T > 0$ is a constant. From this we deduce the inequality

$$\delta(G, \tau J) \leq \frac{\log \rho(G)}{\log \tau^{-1}}.$$

If $\rho(G) = 1$, i. e., if v_p grows polynomially, we conclude that the Hausdorff dimension of the set G^∞ is zero, and the lemma trivially holds. In what follows, we assume that $\rho(G) > 1$. We now show that

$$\delta(G, \tau J) \geq \frac{\log \rho(G)}{\log \tau^{-1}}.$$

We split into two cases:

1. Assume that G is strongly connected. In that case, $v_p \approx T\rho(G)^p$ for $p \gg 1$. Consider a finite ϵ-cover $\mathcal{A}$:

$$\bigcup_{a^j \in \mathcal{A}} D(a^j, \epsilon_j) = G^\infty, \quad \tau^{\kappa_j+1} < \epsilon_j \leq \tau^{\kappa_j} \leq \epsilon, \quad a^j \in \mathcal{A}.$$

Assume that $p \gg 1$. Fix any set $\mathcal{B}_p$. Since $\bigcup_{a^j \in \mathcal{A}} D(a^j, \epsilon_j) \supseteq \mathcal{B}_p$, we deduce that

$$T\rho(G)^p \approx v_p = |\mathcal{B}_p| \leq \sum_{a^j \in \mathcal{A}} |D(a^j, \epsilon_j) \cap \mathcal{B}_p|.$$

Next, observe that the ball $D(a^j, \epsilon_j)$ contains at most $T'\rho(G)^{p-\kappa_j-1}$ points in $\mathcal{B}_p$ for some $0 < T'$. Hence

$$\sum_{a^j \in \mathcal{A}} \rho(G)^{p-\kappa_j-1} \geq T_1\rho(G)^p, \quad T_1 > 0,$$

that is,

$$\sum_{a^j \in \mathcal{A}} \rho(G)^{-\kappa_j-1} \geq T_1, \quad T_1 > 0.$$

It then follows that

$$\Lambda_\alpha(G_\infty) \geq T_1, \quad \alpha \leq \frac{\log \rho(G)}{\log \tau^{-1}}.$$

Hence $\delta(G^\infty, \tau J) \geq \frac{\log \rho(G)}{\log \tau^{-1}}$. This proves the lemma for a strongly connected G.

2. Assume that G is not strongly connected. Since $\rho(G) > 0$, there exists an induced subgraph $G_1 = (V \times V \cap G)$, where $V \subset [n]$, such that G_1 is strongly connected and $\rho(G_1) = \rho(G)$. Clearly, $G_1^\infty \subseteq G^\infty$. Therefore

$$\delta(G, \tau J) \geq \delta(G_1, \tau J) = \frac{\log \rho(G_1)}{\log \tau^{-1}} = \frac{\log \rho(G)}{\log \tau^{-1}}.$$

So the proof is complete. □

We wrap up this section by computing the exact value of $\delta(G, B)$, where B is a nonnegative matrix with the G-cyclic property.

Theorem 7.5. *Let $G \subset [n] \times [n]$ and assume that $\rho(G) > 1$. Let $B = (b_{ij})_{1 \le i, j \le n}$ be a nonnegative matrix with $G(B) \supset G$ that satisfies the G-cycle property. Assume that there exists $0 < \tau < 1$ such that*

$$b_{ij} = \tau^{m_{ij}}, \quad (i, j) \in G, \quad m_{ij} \in \mathbb{Z}.$$

Let $A = (a_{ij})_{1 \le i, j \le n}$ be a matrix of the form (7.2.1). (If $m_{ij} \ge 1$ for $(i, j) \in G$, let $A = B$ and $t = \tau$.) Let G' be the following digraph obtained from G: on each directed edge (i, j), insert $n_{ij} - 1$ new vertices so that the edge (i, j) becomes a directed path of length n_{ij} and no other edges are inserted. Let d_A and d_{tJ} be metrics on G^∞ and G'^∞, respectively. Then G^∞ embeds isometrically into G'^∞. Moreover, G'^∞ is a disjoint union of a finite number of (conformal) copies of G^∞. In particular,

$$\delta(G, B) = \delta(G, A) = \delta(G', tJ) = \frac{\log \rho(G')}{\log t^{-1}}. \tag{7.3.1}$$

Finally, there exists a digraph G'' such that

$$\delta(G, B) = \frac{\log \rho(G'')}{\log \tau^{-1}}.$$

Proof. First, assume that $m_{ij} \ge 1$ for all $(i, j) \in G$. In this case, set $A = B$, $t = \tau$, and $n_{ij} = m_{ij}$ for all $(i, j) \in G$. Any sequence $a = (a_i)_{i \ge 1} \in G^*$ represents an infinite walk in G starting at the vertex a_1. Extend each edge (a_i, a_{i+1}) in G to a chain of length m_{ij} in G'. This extension results in a unique infinite walk in G', defining a map $\phi : G^\infty \to G'^\infty$. Given that the metrics on these spaces are d_B and d_{tJ} and that $b_{ij} = \tau^{m_{ij}}$ for $(i, j) \in G$, it follows that ϕ is an isometry.

For a vertex $v \in G'$ with $v \notin [n]$, define

$$G'^\infty(v) = \{c : (c_i)_{i \ge 1} \in G'^\infty, \text{ with } c_1 = v\}.$$

In other words, $G'^\infty(v)$ is the set of all walks in G' that start at v. Let $p > 1$ be the smallest index such that $c_p \in [n]$. Thus the walk in $c \in G'^\infty(v)$ is the chain $(c_i)_{i \le p}$ and an infinite walk, which is an element of $\phi(G^\infty)$ starting at c_p. The map $\phi_v : G^\infty \to G'^\infty(v)$ is bijective. Therefore $\delta(G^\infty, B) = \delta(G'^\infty, tJ)$, and the theorem follows from Lemma 7.4.

Now consider the case where $b_{ij} \ge 1$ for some $(i, j) \in G$. By applying Lemma 7.3 we can find a matrix A with the required properties. We then use the above arguments for A to establish (7.3.1). The proof for $\delta(G, B) = \frac{\log \rho(G'')}{\log \tau^{-1}}$ is left to the reader. (See Problem 7.6-3.) □

7.4 Limit set on trees

i

Infinite trees. We say that $\mathcal{T} = (V, E)$ is an *infinite tree* if the following conditions are satisfied:

1. V is an infinite set of vertices.
2. $E \subset V \times V$ is a set of undirected edges with no loops, meaning that there are no edges of the form (v, v) for $v \in V$.
3. $\mathcal{T}$ is connected.
4. $\mathcal{T}$ has no cycles.

The tree $\mathcal{T}$ is termed *locally finite* if every vertex $x \in V$ has a finite degree, denoted $\deg(x)$. For any two vertices $x, y \in V$, there exists a unique path (geodesic) connecting x and y. The number of edges in this path is denoted $d_g(x, y)$, which represents the graph distance between x and y.

Fix a reference point $o \in V$ and consider an infinite chain C_1 starting from o. This chain forms an *infinite geodesic*, whose endpoint corresponds to a unique boundary point $c_1 \in \partial \mathcal{T}$. Let C_2 be another geodesic starting at o, corresponding to another boundary point $c_2 \in \partial \mathcal{T}$. Assume that $c_1 \neq c_2$. The chordal distance $d_c(c_1, c_2)$ is defined as follows: the geodesics C_1 and C_2 share a maximal common chain $o - x_1 - \cdots - x_m$. Then the *chordal distance* is given by

$$d_c(c_1, c_2) = \tau^m$$

for a fixed $0 < \tau < 1$.

Automorphism group of a tree. Given an infinite tree $\mathcal{T} = (V, E)$:

- A bijection $\phi : V \to V$ is called a *tree isomorphism* if for all pairs $(x, y) \in V \times V$, the following condition holds for all $x, y \in V$:

$$(\phi(x), \phi(y)) \in E \iff (x, y) \in E. \tag{7.4.1}$$

- Another way to express (7.4.1) is that ϕ must be an isometry, meaning that $d_g(x, y) = d_g(\phi(x), \phi(y))$.
- The bijection ϕ is said to *act freely* if $\phi(x) \neq x$ for all $x \in V$.
- We denote by $\mathrm{Aut}(\mathcal{T})$ the group of isometries of $\mathcal{T}$. (Note that it is easily verified that $\mathrm{Aut}(\mathcal{T})$, together with composition, forms a group. (Problem 7.6-4))

Let $H \subset \mathrm{Aut}(\mathcal{T})$ be a finitely generated free group that acts freely on $\mathcal{T}$ (i. e., each $\phi \in H$, $\phi \neq e$, acts freely on $\mathcal{T}$). Assume that $\gamma_1, \ldots, \gamma_k$ are generators of H. We now construct $\Lambda(H) \subset \partial \mathcal{T}$, the limit set of H, using the generators $\gamma_1, \ldots, \gamma_k$ as follows. Define

$$\Delta = \{(i, j) : i, j \in [2k], |i - j| \neq k\} \subset [2k] \times [2k].$$

Here the generator γ_i corresponds to the symbol i for $i = 1, \ldots, k$, whereas γ_i^{-1} corresponds to the symbol $i + k$ for $i = 1, \ldots, k$. Set

$$\gamma(j) = \gamma_j \quad \text{for } j = 1, \ldots, k, \quad \gamma(j) = \gamma_{j-k}^{-1} \quad \text{for } j = k+1, \ldots, 2k.$$

Let $a = (a_i)_1^\infty \in \Delta^\infty$. Then a induces the following infinite reduced word:

$$w(a) = \gamma(a_1) \cdots \gamma(a_i) \cdots.$$

Let $a^q = (a_i)_1^q \in \Delta^q$, and set

$$w(a^q) = \gamma(a_1) \cdots \gamma(a_q)$$

to be the reduced word of length q corresponding to a^q. We now want to associate with any $a \in \Delta^\infty$ a unique point $c(a) \in \partial T$. A natural approach is the following: Let $C(a^q)$ be the unique chain in T that joins o and $w(a^q)o$. If $C(a^q)$ converges to a unique infinite chain $C(a)$ starting at o, then $c(a)$ is identified with (the endpoint of) $C(a)$. We claim that this is the case where T is locally finite.

Lemma 7.6. *Let T be an infinite, locally finite tree. Suppose $H \subset \text{Aut}(T)$ is a free group with k generators $\gamma_1, \ldots, \gamma_k$ that acts freely on T. For any given $1 \le s \in \mathbb{Z}_+$, there exists $t = t(s) \in \mathbb{Z}_+$ such that the following holds: Let b be any reduced word formed from $\gamma_1, \ldots, \gamma_k$ of length $r > t(s)$. Write $b = ab'$, where a is a reduced word of length $t(s)$, and b' is a reduced word of length $r - t(s)$. Denote by $C(b)$ and $C(a)$ the chains connecting o to bo and ao, respectively. Then $C(b)$ and $C(a)$ share a common subchain of at least length s starting from o.*

In particular, for any sequence $a = (a_i)_1^\infty \in \Delta^\infty$, the sequence of chains $\{C((a_i)_1^q)\}_{q=1}^\infty$ converges to a unique infinite chain $C(a)$ starting from o.

Proof. Define $D(r, o) = \{v \in V : dg(v, o) < r\}$. Since T is locally finite, $D(r, o)$ is a finite set for every $r > 0$. Let

$$U(o) = Ho \subset V, \quad U(r, o) = U(o) \cap D(r, o), \quad H(r, o) = \{h \in H : ho \in U(r, o)\}.$$

Since $U(r, o)$ is finite and H acts freely, $H(r, o)$ must also be finite. Denote by $\mu(r, o)$ the maximum length of all elements in $H(r, o)$, expressed as reduced words in the generators $\gamma_1, \ldots, \gamma_k$. Define

$$D(o) = \max_{1 \le i \le k} dg(\gamma_i o, o).$$

We state that the lemma holds for $t(s) = \mu(D(o) + s, o) + 1$, where $s \ge 1$. We prove the first claim of the lemma by induction on the length of b. Suppose the length of b is $\mu(D(o) + s, o) + 2$. The chain $C(b)$ is constructed as follows: first, consider the chain $C(a)$ from o to ao. Then consider the chain C' from ao to $a(b'o)$. The chain $C(b)$ is constructed by joining $C(a)$ to C' and eliminating any edges traversed twice. Since $b' = \gamma(i)$ with $i \in [2k]$, and H is a group of isometries of $T(H)$, the length of C' is at most $D(o)$. Given that the length of a is $\mu(D(o) + s, o) + 1$, it follows that $dg(ao, o) \ge D(o) + s$. Consequently, when $C(a)$ and C' are joined, at most $D(o)$ edges may be canceled. Thus $C(a)$ and $C(b)$ share a common chain of at least length s starting from o.

Now, assume the lemma holds for all c whose length is $\mu(D(o) + s, o) + m$, where $m \geq 2$. Let b be a reduced word of length $\mu(D(o) + s, o) + m + 1$. Write $b = cb''$, where c is a reduced word of length $\mu(D(o) + s, o) + m + 1$, and b'' is a word of length 1. By the previous arguments, $C(b)$ and $C(c)$ share at least a common chain of length s starting from o. The induction hypothesis implies that $C(a)$ and $C(c)$ have a common chain of length at least s starting from o. Therefore $C(b)$ and $C(a)$ must have a common chain of length at least s starting from o. This completes the proof of the first part of the lemma.

To prove the second part of the lemma, let $a \in \Delta^\infty$. Fix $k \geq 1$, and let $x_k \in C(a^{t(k)+1})$ with $d(x_k, o) = k$. From the earlier arguments, it follows that $C(a)$ is the chain

$$o - x_1 - \cdots - x_k - \cdots. \qquad \square$$

Corollary 7.7. *Let the assumptions of Lemma 7.6 hold. Then $\Lambda(H)$ is isomorphic to Δ^∞, and the metric on $\Lambda(H) \subset \partial T$ induces a metric on Δ^∞.*

Proof. The proof is left as Problem 7.6-5. $\qquad \square$

Example 7.8. Let H be a free group on n generators $a_1, \ldots, a_n$ with the identity e. We then associate with H the following Cayley tree:

$$T(H) = (V, E), \quad V = H, \quad E = \{(g, h) : g^{-1}h \in \{a_1, a_1^{-1}, \ldots, a_n, a_n^{-1}\}\}.$$

For $h \in H$, let $l(h)$ be the length of h as a reduced word in the generators $a_1, \ldots, a_n$ (with $l(e) = 0$). It then follows that

$$dg(g, h) = l(g^{-1}h).$$

Note that $\deg(v) = 2n$, $v \in V$. Then H acts on $T(H)$ as a group of isometries by letting $h(v) = hv$, $h \in H$, $v \in V$. Observe that H acts freely on $T(H)$. Thus any finitely generated subgroup $H' \subset H$ is a finitely generated free subgroup of $\mathrm{Aut}(T(H))$ that acts freely. Hence the conditions of Lemma 7.6 hold. The chordal metric on $\partial T(H)$ induces a metric on $\Lambda(H')$, which, in turn, can be viewed as a metric on Δ^∞ according to Corollary 7.7. It is possible to find the Hausdorff dimension of $\Lambda(H')$ using Theorem 7.5 under the following assumptions. Assume that H' is freely generated by $y_1, \ldots, y_k$ and let $y(i)$, $i = 1, \ldots, 2k$, be defined as above. The quantity

$$\mathrm{can}(y(i), y(j)) = \frac{l(y(i)) + l(y(j)) - l(y(i)y(j))}{2}, \quad 1 \leq i, j \leq 2k, \quad |i - j| \neq k,$$

is the length of the cancellation word in the product $y(i)y(j)$, that is,

$$y(i) = \mu(i, j)\omega(i, j) = \omega(i, j)^{-1}v(i, j),$$

$$l(\omega(i,j)) = \mathrm{can}(\gamma(i),\gamma(j)),$$
$$l(\gamma(i)) = l(\mu(i,j)) + l(\omega(i,j)) = l(\nu(i,j)) + l(\omega(i,j)),$$
$$l(\gamma(i)\gamma(j)) = l(\mu(i,j)) + l(\nu(i,j)).$$

Theorem 7.9. *Let H be a finitely generated free group with the associated Cayley tree $T(H)$. Let $H' \subset H$ be a free subgroup with generators $\gamma_1, \ldots, \gamma_k$. Identify the limit set $\Lambda(H') \subset \partial T(H)$ with Δ^∞ and let $d : \Delta^\infty \times \Delta^\infty \to \mathbb{R}_+$ be the metric induced by the chordal metric on $\Lambda(H')$. Suppose the following condition is satisfied:*

$$\mathrm{can}(\gamma(i,j)) + \mathrm{can}(\gamma(j,l)) < l(\gamma(j)), \quad 1 \le i,j,l \le k, \quad |i-j| \ne k, \quad |j-l| \ne k.$$

Under this condition, the metric d on Δ^∞ is equivalent to the pseudometric d_B, where

$$B = (b_{ij})_{1 \le i,j \le 2k},$$
$$b_{ij} = \tau^{l(\gamma(i)) - 2\mathrm{can}(\gamma(i),\gamma(j))} \quad for \; |i-j| \ne k, \quad 0 < \tau < 1,$$
$$b_{pq} = 1 \quad for \; |p-q| = k.$$

In particular, the Hausdorff dimension $\delta(\Lambda(H'))$ of the limit set $\Lambda(H')$ with respect to the chordal metric d is equal to $\delta(\Gamma, B)$ as given by Theorem 7.5.

Proof. The assumptions of the lemma imply the following. Consider the product $\gamma(i)\gamma(j)\gamma(l)$. The cancellations in $\gamma(j)$ caused by $\gamma(i)$ from the left and $\gamma(l)$ from the right are effectively separated. That is,

$$\gamma(j) = \omega^{-1}(i,j)\kappa(i,j,l)\omega(j,l),$$
$$\nu(i,j) = \kappa(i,j,l)\omega(j,l),$$
$$\mu(j,l) = \omega(i,j)^{-1}\kappa(i,j,l),$$
$$l(\gamma(j)) = l(\omega(i,j)) + l(\kappa(i,j,l)) + l(\omega(j,l)),$$
$$l(\kappa(i,j,l)) > 0.$$

Let $a^q = (a_i)_1^q \in \Delta^q$ and consider the word $w(a^q)$. It follows that

$$l(w(a^q)) = l(\gamma(a_q)) + \sum_{i=1}^{q-1} l(\gamma(a_i)) - 2c(\gamma(a_i),\gamma(a_{i+1})). \tag{7.4.2}$$

Let $a = (a_i)_1^\infty$ and $b = (b_i)_1^\infty$ be elements of Δ^∞ such that $a_i = b_i$ for $i = 1, \ldots, p$ and $a_{p+1} \ne b_{p+1}$. According to Lemma 7.6, the chordal distance between a and b is given by

$$d(a,b) = \tau^{\frac{l(w(a^q)^{-1}w(b^q)) - l(w(a^q)) - l(w(b^q))}{2}}$$

for a sufficiently large q. (In fact, under our assumptions, $q = p + 1$ suffices.)

It follows from (7.4.2), along with the definitions of B and d_B, that:

$$\frac{d_B(a,b)}{K} \le d(a,b) \le Kd_B(a,b), \quad \text{for } a,b \in \Delta^\infty, \quad K \le \tau^{-6D(o)}.$$

Use the results of Theorem 7.5 to complete the proof. $\qquad\qquad\qquad\qquad$ $\square$

7.5 Worked-out problems

1. Show that the techniques used to prove Theorem 7.9 can be applied to the general case under the conditions of Lemma 7.6. In other words, let the conditions of Lemma 7.6 hold. Identify $\Lambda(H)$ with Δ^∞ and let d be the metric on Δ^∞ induced by the chordal metric on $\partial\mathcal{T}$. Define $m = t(D(o)) + 1$, where $t(s)$ and $D(o)$ are as specified in the proof of Lemma 7.6.
Suppose

$$a = (a_i)_{i=1}^\infty, \quad b = (b_i)_{i=1}^\infty \in \Delta^\infty,$$

with $a_i = b_i$ for $i = 1,\ldots,m$. Then show that there exists an integer $\beta((a_i)_1^m)$, which depends only on $a^m = (a_i)_1^m \in \Delta^m$, such that

$$d(a,b) = \tau^{\beta((a_i)_1^m)} d((a_i)_2^\infty, (b_i)_2^\infty).$$

Solution. Let $a' = (a_i)_2^\infty$ and $b' = (b_i)_2^\infty$. By Lemma 7.6 and the arguments in the proof of Theorem 7.9, for sufficiently large q, we have

$$d(a,b) = \tau^{\frac{dg(w((a_i)_1^q)o,w((b_i)_1^q)o)-dg(w((a_i)_1^q)o,o)-dg(w((b_i)_1^q)o,o)}{2}},$$

$$d(a',b') = \tau^{\frac{dg(w((a_i)_2^q)o,w((b_i)_2^q)o)-dg(w((a_i)_2^q)o,o)-dg(w((b_i)_2^q)o,o)}{2}}.$$

Define

$$2\beta = dg(w((a_i)_1^q)o, w((b_i)_1^q)o) - dg(w((a_i)_1^q)o, o) - dg(w((b_i)_1^q)o, o)$$
$$- dg(w((a_i)_2^q)o, w((b_i)_2^q)o) + dg(w((a_i)_2^q)o, o) + dg(w((b_i)_2^q)o, o).$$

We claim that β depends only on $(a_i)_1^m$. Since H is a group of isometries, it follows that

$$dg(w((a_i)_1^q)o, w((b_i)_1^q)o) = dg(w((a_i)_2^q)o, w((b_i)_2^q)o).$$

Define

$$\beta_1 = dg(w((a_i)_1^q)o, o) - dg(w((a_i)_2^q)o, o),$$
$$\beta_2 = dg(w((b_i)_1^q)o, o) - dg(w((b_i)_2^q)o, o).$$

We claim that both β_1 and β_2 depend only on $(a_i)_1^m$.

First, consider β_1. Note that

$$dg(w((a_i)_1^q)o, o) = dg(w((a_i)_2^q)o, \gamma(a_1)^{-1}o).$$

The chain $C((a_i)_2^q)$ connecting o to $w((a_i)_2^q)o$ is composed of the chain $C(a_1 + k)$ from o to $\gamma(a_1+k)o$ plus the chain C' from $\gamma(a_1+k)o$ to $w((a_i)_2^q)o$. (Here $a_1 + k$ is considered modulo $2k$.) Let can denote the length of the portion of the chain that we traverse twice—first, along $C(a_1 + k)$ and then along C'.

Thus we have

$$\beta_1 = 2 \cdot \text{can} - dg(\gamma(a_1)o, o).$$

It remains to show that can depends only on the vector $(a_i)_1^m$. Specifically, can can be computed as follows: Let $C'' = C((a_i)_2^q) \cap C(a_1 + k)$ be the intersection of the two chains. Then can is given by $dg(\gamma(a_1)o, o)$, the length of C''. To show that $C'' = C((a_i)_2^m) \cap C(a_1 + k)$, observe that the length of the word $(a_i)_2^m$ is $t(D(o))$. Therefore, for $q > m$, by Lemma 7.6, $C((a_i)_2^q)$ and $C((a_i)_2^m)$ have a common chain of length $D(o)$ starting from o. By definition, $D(o) \geq dg(\gamma(a_1)o, o)$. Thus $C'' = C((a_i)_2^m) \cap C(a_1 + k)$, and β_1 depends only on $(a_i)_1^m$. Similarly, β_2 depends only on $(a_i)_1^m$.

7.6 Problems

1. Let $B = (b_{ij})_{i,j=1}^n$ be a nonnegative matrix. Define $G(B)$ as the digraph induced by B, where $(i,j) \in G(B)$ if and only if $b_{ij} > 0$. Assume that $b_{ij} < 1$ for all $i, j = 1, \ldots, n$.
   (a) Prove that B satisfies the G-cycle property for every G with $E(G) \subset [n] \times [n]$.
   (b) If $G \subset G(B)$, then verify that d_B is a metric on G^∞.
   (c) Prove that G^∞ is compact with respect to d_B.

2. Show that if two nonnegative matrices A and B are diagonally similar, then $A(\sigma) = B(\sigma)$ for all cycles σ in $[n] \times [n]$.

3. Complete the proof of Theorem 7.5 by proving the existence of a digraph G'' for which

$$\delta(G, B) = \frac{\log \rho(G'')}{\log \tau^{-1}}$$

   as follows:
   - Notice that $t = \tau^{1/L}$, where $L \in \mathbb{N}$.
   - Construct G'' so that $\rho(G'') = \rho(G')^L$. (This can be achieved using tensor products of matrices; $G'' = G(\otimes^L A(G'))$).

4. Given an infinite tree $\mathcal{T}$, prove that $\text{Aut}(\mathcal{T})$ together with composition forms a group.

5. Assume that $\mathcal{T}$ is a locally finite tree and $H \subset \text{Aut}(\mathcal{T})$ is a finitely generated free group that acts freely on $\mathcal{T}$. Then the proof of Lemma 7.6 shows that the limit set $\Lambda(H) \subset \partial\mathcal{T}$ corresponds to all infinite geodesics originating from o such that each geodesic intersects infinitely many elements of Ho. Consequently, the definition of $\Lambda(H)$ is independent of the choice of generators for H. Fix a set of free generators $\gamma_1, \ldots, \gamma_k$. Assume that the assumptions of Lemma 7.6 hold. Use Lemma 7.6 to prove Corollary 7.7.

A Appendices

Outline. This chapter comprises ten sections, labeled from A.1 to A.10, where we succinctly cover basic results in set theory, elementary number theory, mathematical reasoning, mathematical analysis, abstract algebra, linear algebra, and topology.

A.1 Basic set theory and functions

In this section, we briefly review fundamental results on sets and related operators, equivalence relations on sets, and the concepts of countability and uncountability. We also discuss functions and their specific families such as surjections and injections, bijections, and permutations. For a detailed treatment, see [39].

Sets. A *set* is defined as an unordered collection of distinct objects, referred to as the elements or members of the set. Membership in or exclusion from a set is indicated using the symbols $\in$ and $\notin$, respectively. The *empty set* is the set containing no elements. The *intersection* of two sets A and B, denoted as $A \cap B$, is the set of elements common to both A and B. The *union* of these sets, denoted by $A \cup B$, includes all elements that appear in at least one of the sets. Sets A and B are considered *equal* (written $A = B$) if they contain precisely the same elements. If every element of A is also an element of B, then A is a *subset* of B, denoted $A \subset B$. Two sets are *disjoint* if they have no elements in common. The *set difference* $B \setminus A$ is formed by removing all elements of a set A from a set B. The *power set* $\mathcal{P}(A)$ of a set A is defined as the set of all subsets of A, including both the empty set and the set A itself.

The symbols $a, b, \ldots$ are known as *indices*, and the set $I = \{a, b, c, \ldots, z\}$ is referred to as the *index set*. The collection of sets W_a through W_z constitutes a *family of sets*, as they are related by a common definition. Each index indicates a specific set within this family.

For two sets A and B, the *Cartesian product* $A \times B$ is the set of all ordered pairs (a, b) where $a \in A$ and $b \in B$:

$$A \times B = \{(a, b) \mid a \in A \text{ and } b \in B\}.$$

A *relation* from a set A to a set B is defined as a subset of $A \times B$.

Equivalence relations. A *relation on a set* A is a relation from A to itself, i.e., a subset of $A \times A$. For a relation R on A (where $R \subset A \times A$), we denote $x \sim y$ if $(x, y) \in R$.

An *equivalence relation* on a set A is a relation that satisfies the following properties:

1. *Reflexivity*: For all $x \in A$, $x \sim x$.
2. *Symmetry*: For all $x, y \in A$, if $x \sim y$, then $y \sim x$.
3. *Transitivity*: For all $x, y, z \in A$, if $x \sim y$ and $y \sim z$, then $x \sim z$.

https://doi.org/10.1515/9783111337388-008

Given an equivalence relation $\sim$ on A and an element $x \in A$, the set

$$E_x = \{y \in A \mid x \sim y\}$$

is called the *equivalence class* of x. Another notation for the equivalence class E_x is $[x]$.

A collection of nonempty subsets $A_1, A_2, \ldots$ of A is called a *partition* of A if it satisfies the following conditions: 1. $A_i \cap A_j = \emptyset$ for $i \neq j$; 2. $\bigcup_i A_i = A$.

The following fundamental results describe the relationship between equivalence relations and partitions of a set. Detailed proofs can be found in [39].

Theorem A.10. *Given an equivalence relation on a set A, the set of distinct equivalence classes forms a partition of A.*

Theorem A.11. *Given a partition of A into sets $A_1, \ldots, A_n$, the relation defined by "$x \sim y$ if and only if x and y belong to the same set A_i from the partition" is an equivalence relation on A.*

Functions. Let X and Y be two sets. A *function* $f : X \to Y$, denoted by $x \mapsto f(x)$, is defined so that for each $x \in X$, $f(x)$ corresponds to exactly one element in Y. In this context, X is called the *domain* of the function f, and Y is referred to as the *codomain*. The notation $f(x)$ indicates that the rule f is applied to the element $x \in X$. The element $y = f(x)$ is referred to as the *image* of x under the function f.

If $A \subset X$, then the set $f(A)$, called the *image of A* under f, is defined as

$$f(A) = \{y \in Y \mid \exists x((x \in A) \wedge (f(x) = y))\}.$$

If $A = X$, then $f(X)$ is referred to as the *image of the function f*, and we denote it as $f(X) = \operatorname{Im} f$.

For any $y \in Y$, we define $f^{-1}(y) = \{x \in X \mid f(x) = y\}$. In general, for $B \subset Y$, we define $f^{-1}(B) = \{x \in X \mid f(x) \in B\}$. The set $f^{-1}(B)$ is called the *preimage* of B under f. It is evident that $f^{-1}(Y) = X$. The function $f : X \to Y$ is called *injective* if for any two elements $x_1, x_2 \in X$, the condition $x_1 \neq x_2$ implies $f(x_1) \neq f(x_2)$.

The function $f : X \to Y$ is called *surjective* (or onto) if $f(X) = Y$.

The function $f : X \to Y$ is referred to as *bijective* (or one-to-one) if it is both injective and subjective.

Countable/uncountable sets. Two sets A and B are said to have the same *cardinality*, denoted as $|A| = |B|$, if there exists a bijection $f : A \to B$. If $|A| = |B|$, then A and B are sometimes called *equinumerous*, written as $A \sim B$. If no such bijective function exists, then the sets have different cardinalities, meaning that $|A| \neq |B|$.

A set A is called *finite* if it is either empty or there exists a natural number $n \in \mathbb{N}$ such that there is a bijection $f : \{1, \ldots, n\} \to A$. Otherwise, the set A is classified as *infinite*. A set A is called *countably infinite* if $A \sim \mathbb{N}$. We say that A is *countable* if either $A \sim \mathbb{N}$ or A is finite. If no bijection exists between $\mathbb{N}$ and A, then A is referred to as *uncountable*.

It is worth mentioning that:

- There is no surjection from a set A to its power set $P(A)$. Therefore A and $P(A)$ cannot be equinumerous.
- Any subset of a countable set is also countable.
- If a set I is countable and A_i is countable for every $i \in I$, then the union $\bigcup_{i \in I} A_i$ is countable.
- the set of real numbers $\mathbb{R}$ is uncountable.

Permutations. Given a set A, a *permutation* of A is a function $f : A \rightarrow A$ that is both subjective and injective. A *permutation group* of A is a set of permutations of A that forms a group under function composition. We will focus specifically on the case where $A = [n]$ for some fixed integer n. This means that each group element will permute this set. The *symmetric group* S_n is the group of all permutations of the set $[n]$. For example, the group S_3 consists of six elements. There are six permutations because there are three choices for where to send 1, two choices for where to send 2, and one choice for where to send 3.

A.2 Euclid's algorithm

We briefly overview the Euclid's algorithm. The reader is referred to [42] for more detail.

The *greatest common divisor* (gcd) of two integers a and b, which cannot both be zero, is defined as a positive common divisor $d > 0$ of a and b such that every other common divisor of a and b divides d. We denote the greatest common divisor of a and b by $\gcd(a, b)$. For convenience, we occasionally define $\gcd(0, 0) = 0$.

Notation: We write $d \mid a$ to signify that d is a divisor of a. We express $a \perp b$ to indicate that the integers a and b are *coprime*, meaning that $\gcd(a, b) = 1$.

Euclid's algorithm. Euclid's algorithm is a technique for determining the greatest common divisor of two positive integers a and b. The algorithm is based on the observation that any common divisor d of a and b must also divide the difference $a - b$. Specifically, if $a = a'd$ and $b = b'd$ for some integers a' and b', then $a - b = (a' - b')d$, indicating that d divides $a - b$.

Algorithm: Subtractive Algorithm for Greatest Common Divisors
Input: Positive integers a, b
Output: $d = \gcd(a, b)$

$$\text{while } (a \neq b) \text{ do}$$
$$(a, b) = (\min(a, b), \max(a, b) - \min(a, b));$$
$$\text{return } a;$$

Note that we use the compound assignment $(a, b) = (c, d)$, which simultaneously assigns the values of c and d to a and b, respectively. An invariant of the algorithm is the relation:

$$gcd(a, b) = gcd(\min(a, b), \max(a, b) - \min(a, b)).$$

The algorithm terminates because the values of a and b decrease in each iteration while remaining positive, ensuring that they will converge after a finite number of steps. When the algorithm concludes, the variables a and b will be equal. Throughout the iterations, we have established that the gcd of the two variables remains unchanged. Since $gcd(d, d) = d$ for any nonzero integer d, we can conclude that the algorithm produces the correct result.

i A.3 Principle of induction

We borrow the following definitions and results from [21].

Well-ordering axiom for the integers. If B is a nonempty subset of $\mathbb{Z}$ that is bounded below, meaning that for all $b \in B$, there exists an integer $n \in \mathbb{Z}$ such that $n \leq b$, then B has the smallest element. This means that there exists $b_0 \in B$ such that $b_0 < b$ for all $b \in B$, where $b \neq b_0$.

Theorem A.12 (Well-ordering principle for $\mathbb{N}$). *Every nonempty set of nonnegative integers has a least element.*

Next, we state the induction principle.

Theorem A.13 (Principle of mathematical induction). *Let P be a set of integers such that:*
1. *$a \in P$,*
2. *If an integer $k \geq a$ is in P, then the integer $k + 1 \in P$.*

Then $P = \{x \in \mathbb{Z} \mid x \geq a\}$, which means that P consists of all integers greater than or equal to a.

It is proved that the principles of mathematical induction and well-ordering are logically equivalent.

Theorem A.14 (Principle of strong mathematical induction). *Let P be a set of integers such that:*
- *$a \in P$,*
- *If all integers k with $a \leq k \leq n$ are in P, then the integer $n + 1 \in P$.*

Then $P = \{x \in \mathbb{Z} \mid x \geq a\}$, meaning that P consists of all integers greater than or equal to a.

Note that the principle of strong mathematical induction is equivalent to both the well-ordering principle and the principle of mathematical induction.

A.4 Convex sets

- A set $C \subset \mathbb{R}^n$ is said to be *convex* if for any two points $\mathbf{x}, \mathbf{y} \in C$ and any $0 \leq \lambda \leq 1$, the line segment between them lies entirely within the set. This means that

$$\lambda\mathbf{x} + (1-\lambda)\mathbf{y} \in C.$$

- A *convex combination* of points $\mathbf{x}_1, \ldots, \mathbf{x}_k \in \mathbb{R}^n$ is any point of the form

$$\mathbf{x} = \lambda_1\mathbf{x}_1 + \lambda_2\mathbf{x}_2 + \cdots + \lambda_k\mathbf{x}_k,$$

where $\lambda_1 + \cdots + \lambda_k = 1$, and $\lambda_i \geq 0$ for all i.
- Let

$$\text{conv}(\mathbf{x}_1, \ldots, \mathbf{x}_k) := \left\{ \mathbf{x} \in \mathbb{R}^n : \mathbf{x} = \sum_{l=1}^{k} \lambda_l\mathbf{x}_l \text{ for all } \lambda_1, \ldots, \lambda_k \geq 0 \text{ with } \sum_{l=1}^{k} \lambda_l = 1 \right\}.$$

Then $\text{conv}(\mathbf{x}_1, \ldots, \mathbf{x}_k)$ is called the *convex hull spanned by* $\mathbf{x}_1, \ldots, \mathbf{x}_k$, that is, the convex hull of a set is the set of all convex combinations of points within that set.
- The convex hull is a convex set.

A.5 Newton's binomial theorem

The definition of *n choose k* (often written as $\binom{n}{k}$) represents the number of ways to choose k elements from a set of n distinct elements without regard to the order of selection. This is also known as a *binomial coefficient*. It is defined as

$$\binom{n}{k} = \frac{n!}{k!(n-k)!},$$

where $n!$ is the product of all positive integers up to n, and $k!$ is the product of all positive integers up to k. We may use induction to prove the following theorem due to Newton. See [41] for more information.

Theorem A.15 (Newton's binomial theorem). *Let $x, y \in \mathbb{R}$ and $n \in \mathbb{N}^+$. We have the binomial expansion*

$$(x+y)^n = \sum_{k=0}^{n} \binom{n}{k}x^{n-k}y^k = \binom{n}{0}x^n + \binom{n}{1}x^{n-1}y + \cdots + \binom{n}{n-1}xy^{n-1} + \binom{n}{n}y^n.$$

This can also be expressed as

$$(x+y)^n = \sum_{k=0}^{n} \binom{n}{k}x^k y^{n-k}.$$

A.6 Big O notation

In mathematics, it is essential to understand the concept of approximation error. For example, we often write

$$e^x = 1 + x + \frac{x^2}{2} + O(x^3)$$

to convey that the error is smaller in absolute value than some constant multiple of x^3 when x is sufficiently close to zero.

Formally, suppose f and g are two functions defined on some subset of real numbers. We write

$$f(x) = O(g(x)) \quad \text{as } x \to 0$$

if there exist positive constants e and C such that

$$|f(x)| \leq C|g(x)| \quad \text{for all } |x| < \epsilon.$$

The following is a list of common classes of functions, ordered by their growth rates, frequently encountered in algorithm analysis. Here c is some arbitrary constant:
- $O(1)$ – constant;
- $O(\log(n))$ – logarithmic;
- $O((\log(n))^c)$ – polylogarithmic;
- $O(n)$ – linear;
- $O(n^2)$ – quadratic;
- $O(n^c)$ – polynomial;
- $O(c^n)$ – exponential.

A.7 Basic mathematical analysis

A.7.1 Metric spaces, basic topology, and series

This section provides a concise overview of metric spaces and basic concepts in topology. For a more comprehensive discussion, the reader is encouraged to refer to [37].

A *metric space* (X, d) is a mathematical structure consisting of a set X together with a distance function $d : X \times X \to \mathbb{R}^+$, commonly referred to as a metric, which gives rise to the term "metric space." This function assigns a nonnegative real number $d(x, y)$ to each pair of points $x, y \in X$, representing the distance between them. The distance function must satisfy the following properties:
1. If $d(x, y) = 0$, then $x = y$.
2. For all $x, y \in X$, we have the symmetry: $d(x, y) = d(y, x)$.
3. For all $x, y, z \in X$, we have the triangle inequality: $d(x, z) \leq d(x, y) + d(y, z)$.

In a metric space, we can discuss concepts of convergence and continuity in a manner similar to that in $\mathbb{R}^n$. Let (X, d) be a metric space.

– For a point $x \in X$ and a radius $\epsilon > 0$, the *ball* of radius ϵ centered at x is defined using the distance d as follows:

$$B_d(x, \epsilon) = \{y \in X \mid d(x, y) < \epsilon\}.$$

– A sequence $\{x_n\}_{n \in \mathbb{N}} \subset X$ is said to *converge* to x, denoted $\lim_{n \to \infty} x_n = x$, if for every $\epsilon > 0$, there exists an integer $N > 0$ such that $x_n \in B_d(x, \epsilon)$ for all $n \geq N$. Otherwise, $\{x_n\}_{n \in \mathbb{N}} \subset X$ *diverges*.
– A set $U \subset X$ in a metric space (X, d) is considered *open* if for every $x \in U$, there exists $\epsilon > 0$ such that $B_d(x, \epsilon) \subset U$.
– A set $C \subset X$ is deemed *closed* if its complement $X \setminus C$ is open.
– A function $f : X \to Y$ is called *continuous* if for every $\epsilon > 0$, there exists $\delta > 0$ such that $f(B_d(X, \delta)) \subset B_d(Y, \epsilon)$.
– Let $A \subset X$. A point $x \in X$ is called a *boundary point* of A if every ball centered at x contains both a point in A and a point in $X \setminus A$. The set of all boundary points of A is denoted by $\partial(A)$.
– An *accumulation point* or *limit point* of a set $A \subset X$ is a point $x \in X$ such that every neighborhood of x contains a point in A different from x. The set of all limit points of A is denoted by A'.
– The *closure* of a subset A of a metric space X is the set

$$\overline{A} = A \cup A'.$$

– A subset $K \subset X$ is *compact* if every sequence $\{x_n\}_{n \in \mathbb{N}} \subset K$ contains a convergent subsequence $\{x_{n_k}\}_{k \in \mathbb{N}}$ whose limit $\lim_{k \to \infty} x_{n_k} = x$ is also in K.
– A subset $S \subset X$ is said to be *bounded* if there exists $r > 0$ such that for all points s and t in S, the distance $d(s, t) < r$.
– Closed and bounded subsets of $\mathbb{R}^n$ are compact, as stated in the *Heine–Borel theorem*. Conversely, in $\mathbb{R}$, if a set is not bounded or not closed, it cannot be compact.
– Closed subsets of compact sets are compact.
– If C is closed and K is compact, then $C \cap K$ is compact.
– We say that (X, d) is a *connected metric space* if X cannot be written as a disjoint union $X = Y_1 \cup Y_2$, where Y_1 and Y_2 are nonempty open subsets of X. (A *disjoint union* means that $X = Y_1 \cup Y_2$ and $Y_1 \cap Y_2 = \emptyset$.) A metric space that is not connected is said to be *disconnected*. Similarly, connectedness is defined for any $A \subset X$.
– A set $D \subset X$ is called a *domain* if D is connected and open.
– The *diameter* of a set $A \subset X$ is defined as the supremum of the distances between all pairs of points in A and is commonly denoted by diam(A). Specifically,

$$\mathrm{diam}(A) = \sup\{d(x, y) : x, y \in A\}.$$

Note that A is bounded if its diameter is finite.
- Let X and Y be two metric spaces with metrics d_X and d_Y, respectively. A map

$$f : X \rightarrow Y$$

is called *distance-preserving* if

$$d_Y(f(A), f(B)) = d_X(A, B)$$

for all $A, B \in X$.
A bijective distance-preserving map is called an *isometry*. Two metric spaces are called *isometric* if there exists an isometry from one to the other.
- Given metric spaces X and Y, we say X *embeds isometrically* into Y if there exists an injective isometry from X to Y.

Series. A *series* can be thought of as the sum of a sequence. Given a sequence $\{x_n\}_{n=0}^{\infty}$, the corresponding series is

$$\sum_{n=0}^{\infty} x_n = x_0 + x_1 + x_2 + \cdots.$$

This series is associated with another sequence, known as the sequence of *partial sums*, denoted by $\{s_n\}_{n=0}^{\infty}$, where

$$s_n = \sum_{i=0}^{n} x_i.$$

For instance,

$$s_0 = x_0, \quad s_1 = x_0 + x_1, \quad s_2 = x_0 + x_1 + x_2, \quad \ldots.$$

- A series is said to *converge* if the sequence of partial sums converges; otherwise, the series is considered *divergent*.
- A real-valued series $\sum x_n$ is called *absolutely convergent* if the series of absolute values $\sum |x_n|$ is convergent.

A.7.2 The Cauchy–Schwarz inequality

We state the *Cauchy–Schwarz inequality* for real numbers, the proof of which can be found in [46].

Theorem A.16. *For all* $x_i, y_i \in \mathbb{R}$, $i = 1, \ldots, n$, *we have*

$$\left|\sum_{i=1}^{n} x_i y_i\right| \le \sum_{i=1}^{n} |x_i|\,|y_i| \le \sqrt{\sum_{i=1}^{n} x_i^2}\,\sqrt{\sum_{i=1}^{n} y_i^2}. \tag{A.7.1}$$

Equality holds if there exist $a, b \in \mathbb{R}$ with $a^2 + b^2 > 0$ such that $ax_i = by_i$ for $i = 1, \ldots, n$ ($\mathbf{x} := (x_1, \ldots, x_n)^\top$ *and* $\mathbf{y} := (y_1, \ldots, y_n) \in^\top \in \mathbb{R}^n$ *are* collinear).

A.7.3 Smooth, Lipschitz, and harmonic functions

- Let U be a nonempty open set in $\mathbb{R}^n$ for some positive integer n, and let f be a continuous real- or complex-valued function on U. A function f is said to be *continuously differentiable* on U if the first partial derivatives

$$\frac{\partial f}{\partial x_1}, \ldots, \frac{\partial f}{\partial x_n}$$

exist and are continuous on U. Similarly, if the first partial derivatives of f are also continuously differentiable on U, then f is said to be *twice continuously differentiable* on U.

If k is a positive integer, and all derivatives of f of order up to and including k exist on U and are continuous on U, then f is said to be k *times continuously differentiable* on U.
- If the derivatives of f of all orders exist and are continuous on U, then f is said to be *infinitely differentiable* on U or simply *smooth*.
- We denote by $C^k(U)$ the space of k times continuously differentiable functions on U for a positive integer k. This can be extended to $k = 0$ by letting $C^0(U)$ be the space $C(U)$ of continuous functions on U. Similarly, $C^\infty(U)$ denotes the space of smooth functions on U.
- A real-valued function f on a metric space (X, d) is said to be *L-Lipschitz* if there exists a constant $L \ge 1$ such that

$$|f(x) - f(y)| \le L \cdot d(x, y)$$

for all x and y in X. Naturally, this definition extends beyond real-valued functions. In general, a map $f : X \to Y$ between metric spaces is called Lipschitz or L-Lipschitz if the above condition holds with the constant $L \ge 1$. Note that Lipschitz functions are smooth functions of metric spaces.

Maximum principle. Let D be a domain in $\mathbb{C}$. We consider both complex- and real-valued functions of $z = x + \mathbf{i}y \in D$. The function will be denoted by $h(z)$, or simply h, to represent both a function of the complex variable z and a function of the real variables x and y. The notation h_x indicates the partial derivative of h with respect to x, and, similarly, h_{xx}, h_{yy}, h_{xy}, and h_{yx} denote the second-order partial derivatives of h. A function

$h : D \to \mathbb{R}$ is called *harmonic* on D if $h \in C^2(D)$ and satisfies the equation $h_{xx} + h_{yy} = 0$ on D.

Theorem A.17. *Let h be a harmonic function on a domain D in $\mathbb{C}$. If h attains a maximum in D, then h is constant.*

Implicit function theorem. Suppose $F(x, y)$ is continuously differentiable in a neighborhood of a point $(a, b) \in \mathbb{R}^n \times \mathbb{R}$ and $F(a, b) = 0$. Assume further that $\frac{\partial F}{\partial y}(a, b) \neq 0$. Then there exist $\delta > 0, \varepsilon > 0$, and a box

$$B = \{(x, y) : |x - a| < \delta, |y - b| < \varepsilon\}$$

such that:

1. For each x such that $|x - a| < \delta$, there is a unique y with $|y - b| < \varepsilon$ for which $F(x, y) = 0$. This defines a function f on $\{x : |x - a| < \delta\}$ such that

$$F(x, y) = 0 \iff y = f(x) \quad \text{for } (x, y) \in B.$$

2. The function f is continuous.
3. The function f is continuously differentiable, and

$$Df(x) = -\frac{D_x F(x, f(x))}{\frac{\partial F}{\partial y}(x, f(x))},$$

where

$$Df = \left[\frac{\partial f}{\partial x_1}, \dots, \frac{\partial f}{\partial x_n} \right] \quad \text{and} \quad D_x F = \left[\frac{\partial F}{\partial x_1}, \dots, \frac{\partial F}{\partial x_n} \right].$$

A.7.4 Taylor series

The *Taylor series* of a function is the sum of an infinite series defined as follows. A proof of this theorem can be found in [43].

Theorem A.18 (Taylor series theorem). *Let f be a function that is analytic at $x = a$. Then we can express $f(x)$ as the following power series, known as the Taylor series of $f(x)$ at $x = a$:*

$$f(x) = f(a) + f'(a)(x - a) + \frac{f''(a)}{2!}(x - a)^2 + \frac{f'''(a)}{3!}(x - a)^3 + \cdots.$$

This series is valid for x within a radius of convergence $|x - a| < R$ with $R > 0$, or it may converge for all x.

Denoting the nth derivative of f at $x = a$ as $f^{(n)}(a)$, we can rewrite the series as

$$f(x) = \sum_{n=0}^{\infty} c_n(x - a)^n,$$

where the coefficients are given by

$$c_n = \frac{f^{(n)}(a)}{n!}.$$

The Taylor series for f at $x = 0$ is known as the *Maclaurin series* for f. The coefficients $c_n = \frac{f^{(n)}(0)}{n!}$ are called the *Maclaurin coefficients*.

A.7.5 L'Hôpital's rule

1. The $\frac{0}{0}$ indeterminate case: Suppose that f and g are differentiable functions over an open interval containing a, except possibly at a. If

$$\lim_{x \to a} f(x) = 0 \quad \text{and} \quad \lim_{x \to a} g(x) = 0,$$

then

$$\lim_{x \to a} \frac{f(x)}{g(x)} = \lim_{x \to a} \frac{f'(x)}{g'(x)},$$

provided that the limit on the right exists (including ∞ or $-\infty$.)

2. The $\frac{\infty}{\infty}$ indeterminate case: Suppose that f and g are differentiable functions over an open interval containing a, except possibly at a. If

$$\lim_{x \to a} f(x) = \infty \quad (\text{or} - \infty) \quad \text{and} \quad \lim_{x \to a} g(x) = \infty \quad (\text{or} - \infty),$$

then

$$\lim_{x \to a} \frac{f(x)}{g(x)} = \lim_{x \to a} \frac{f'(x)}{g'(x)},$$

assuming the limit on the right exists (including ∞ or $-\infty$.)

Note that these results hold for one-sided limits and when $a = \infty$ or $a = -\infty$, provided that the limits on the right exist or are infinite.

A.8 Groups and rings

We briefly overview groups and rings. The reader is referred to [23] for more detail.

Basic definitions (groups). Given a nonempty set A, a *binary operation* on A is a function defined from $A \times A$ to A. A *group* G is a nonempty set of elements G, along with a binary operation $\cdot : G \times G \to G$, that together satisfy four fundamental properties:

1. *Closure*: If a and b are two elements in G, then the product $a \cdot b$ is also in G.
2. *Associativity*: The group operation is associative, i. e., for all a, b, c in G, $(a \cdot b) \cdot c = a \cdot (b \cdot c)$.
3. *Identity*: There exists an identity element e (sometimes denoted by 1) such that $e \cdot a = a = a \cdot e$ for every element a in G.
4. *Inverse*: Each element in G has an inverse. For each element a in G, there exists an element $b = a^{-1}$ in G such that $a \cdot a^{-1} = a^{-1} \cdot a = e$.

We denote the group G with the operation "$\cdot$" by $(G, \cdot)$ or simply by G when the operation is clear from the context. Also, we sometimes use the additive notation $(G, +)$.

We now bring up some basic definitions and results concerning groups.

– If G has a finite number of elements, then it is called a *finite group*, and the number of elements is referred to as the *order* of the group. Otherwise, G is called an *infinite group*.

– We say G is *abelian* if $a \cdot b = b \cdot a$ for all $a, b \in G$.

– A nonempty subset of G that is closed under the group operation and contains the inverse of each of its elements is called a *subgroup* of G.

– For all $a \in G$ and $n \in \mathbb{N}$ we define

$$a^n := \underbrace{a \cdot a \cdot \ldots \cdot a}_{n \text{ times}}$$

and

$$a^{-n} = \underbrace{a^{-1} \cdot a^{-1} \cdot \ldots \cdot a^{-1}}_{n \text{ times}}.$$

– The *order* of $a \in G$ is the smallest positive integer n such that $a^n = e$. If there is no such positive integer n, then the order of the element is considered infinite. The order of a is denoted by $\mathrm{ord}(a)$.

– A *cyclic group* is a group G that can be generated by a single element a, known as the *group generator*, that is, $G = \{a^i \mid i \in \mathbb{Z}\}$. In this case, we use the notation $G = \langle a \rangle$. It is worth mentioning that $|\langle a \rangle| = \mathrm{ord}(a)$.

– A subset $X \subset G$ is called a *generating set* for G if every element $g \in G$ can be expressed as a product of powers of elements from X:

$$g = x_1^{n_1} \cdot x_2^{n_2} \cdot \ldots \cdot x_r^{n_r},$$

where $x_i \in X$ and $n_i \in \mathbb{Z}$. We also say that X generates G and write $G = \langle X \rangle$. If G has a finite generating set, we say that G is a *finitely generated group*.

– Let X be a set and G be a group. A *(left) action* of G on X is a function $G \times X \to X$ given by $(g, x) \mapsto gx$, where the following conditions hold:
    1. $e \cdot x = x$ for all $x \in X$, where e is the identity element of G.
    2. $(g_1 \cdot g_2) \cdot x = g_1 \cdot (g_2 \cdot x)$ for all $x \in X$ and all $g_1, g_2 \in G$.

Under these conditions, the set X is called a *G-set*.

Free groups. Let X be an arbitrary set. A word in X is a finite sequence of elements (possibly empty) denoted by w, which we write as $w = y_1 \cdot \ldots \cdot y_n$, where $y_i \in X$. The number n is called the *length of the word w*, denoted by $l(w)$. The empty word is denoted by ε, with the convention that $l(\varepsilon) = 0$.

Consider the set

$$X^{-1} = \{x^{-1} \mid x \in X\},$$

where x^{-1} is a formal symbol derived from x and -1. If $x \in X$, then the symbols x and x^{-1} are called *literals* in X. Define

$$X^{\pm 1} = X \cup X^{-1},$$

the set of all literals in X. For any literal $y \in X^{\pm 1}$, we define y^{-1} as

$$y^{-1} = \begin{cases} x^{-1} & \text{if } y = x \in X, \\ x & \text{if } y = x^{-1} \in X. \end{cases}$$

An expression of the form

$$w = x_{i_1}^{a_1} \ldots x_{i_n}^{a_n} \quad (x_{i_j} \in X, a_j \in \{1, -1\})$$

is called a *group word* in X. Thus a group word in X is simply a word in the alphabet $X^{\pm 1}$.

A group word

$$w = y_1 \ldots y_n \quad (y_i \in X^{\pm 1})$$

is called *reduced* if for all $i \in [1, n-1]$, we have $y_i \neq y_{i+1}^{-1}$0, in other words, w does not contain any *subword* of the form yy^{-1} for any literal $y \in X^{\pm 1}$. We also assume that the empty word is reduced.

Now let G be a group, and let $X \subset G$. Then every group word $w = x_{i_1}^{a_1} \ldots x_{i_n}^{a_n}$ in X determines a unique element from G, equal to the product $x_{i_1}^{a_1} \cdot \ldots \cdot x_{i_n}^{a_n}$ of the elements $x_{i_j}^{a_j} \in G$. In particular, the empty word ε corresponds to the identity element 1 of G.

A group G is called a *free group* if there exists a generating set X of G such that every nonempty reduced group word in X defines a nontrivial element of G. In this case, X is called a *free basis* of G, and G is said to be *free* on X or *freely generated* by X. It follows from this definition that every element of a free group on X can be represented by a reduced group word on X. Moreover, different reduced words on X define different elements in G.

Cayley graphs. Given a group G and a generating set X of G, the *Cayley graph* $\mathrm{Cay}(G, X)$ has the vertex set G. For all $g \in G$ and $x \in X$, there is a directed edge from g to

gx labeled by x. If $x \in X$ has order two, then for each $g \in G$, the directed edges from g to gx and from gx to g (both labeled x) are replaced by a single undirected edge joining g and gx, also labeled x. A Cayley graph is called a *Cayley tree* if the resulting graph is a tree. (See Chapter 2 for the definition of a tree and related concepts.)

Rings. An abelian group $(R, +)$ is called a *ring* if there exists a second operation, called a product, denoted by $a \cdot b$ for any $a, b \in R$, which satisfies the following properties:

1. *Associativity:*

$$(a \cdot b) \cdot c = a \cdot (b \cdot c) \quad \text{for all } a, b, c \in R.$$

2. *Distributivity:*

$$a \cdot (b + c) = a \cdot b + a \cdot c \quad \text{and} \quad (b + c) \cdot a = b \cdot a + c \cdot a \quad \text{for all } a, b, c \in R.$$

3. *Existence of identity:* There exists an element $1 \in R$ such that $1 \cdot a = a \cdot 1 = a$ for all $a \in R$.

Note that:
- The ring $(R, +, \cdot)$ is called a *commutative ring* if $a \cdot b = b \cdot a$ for all $a, b \in R$. Otherwise, R is called a *noncommutative ring*.
- A *subring* of a ring R is a subset S of R that contains the identity element 1 and is closed under subtraction and multiplication.
- A commutative ring R is called a *field* if every nonzero element $a \in R$ has a unique inverse, denoted by a^{-1}, such that $a \cdot a^{-1} = 1$. A field is usually denoted by $\mathbb{F}$.

A.9 Linear algebra and matrices

We cover some standard terminologies and fundamental results concerning vector spaces, matrices, and their spectral properties along with inner product spaces and norms. We also include a brief discussion about the Gram–Schmidt process. For a basic introduction to these topics, we may refer to [16], and for a more in-depth treatment, [15] is recommended.

A.9.1 Vector spaces

An abelian group $(V, +)$ is called a *vector space* over a field $\mathbb{F}$ if for all $a \in \mathbb{F}$ and $v \in V$, the product av (referred to as scalar multiplication) is an element of V, and this operation satisfies the following properties for all $a, b \in \mathbb{F}$ and $u, v \in V$:
1. $a(u + v) = au + av$;
2. $(a + b)v = av + bv$;

3. $(ab)v = a(bv)$;
4. $1v = v$.

The vector space V over $\mathbb{F}$ is also known as an $\mathbb{F}$-vector space.

We quickly overview a couple of main concepts and results about vector spaces:

−   A nonempty subset $U \subset V$ is called a *subspace* of V if for all $u_1, u_2 \in U$ and $a_1, a_2 \in \mathbb{F}$, we have $a_1 u_1 + a_2 u_2 \in U$. The subspace $U = \{0\}$ is called the *zero* or *trivial subspace* of V.

−   The elements of V are referred to as *vectors*.

−   Consider vectors $x_1, \ldots, x_n \in V$. Then, for any scalars $a_1, \ldots, a_n \in \mathbb{F}$, the expression $a_1 x_1 + \cdots + a_n x_n$ is called a *linear combination* of $x_1, \ldots, x_n$. The expression $0 = \sum_{i=1}^{n} 0 x_i$ represents the *trivial combination*.

−   The vectors $x_1, \ldots, x_n$ are said to be *linearly independent* if the equation $0 = \sum_{i=1}^{n} a_i x_i$ implies that $a_1 = \cdots = a_n = 0$. Otherwise, the vectors are *linearly dependent*.

−   A single vector x is linearly independent if and only if $x \neq 0$.

−   The set of all linear combinations of $x_1, \ldots, x_n$ is called the *span* of $x_1, \ldots, x_n$, denoted by span$\{x_1, \ldots, x_n\}$. Specifically,

$$\text{span}\{x_1, \ldots, x_n\} = \left\{ \sum_{i=1}^{n} a_i x_i \mid a_i \in \mathbb{F} \right\}.$$

In particular, we use the notation span(x) to denote span$\{x\}$ for $x \in V$. Clearly, span$\{x_1, \ldots, x_n\}$ is a subspace of V. (Why?) More generally, the span of a set $S \subset V$ is the set of all linear combinations of elements of S, that is, the set of all linear combinations of all finite subsets of S.

−   A vector space V over a field $\mathbb{F}$ is termed *finite-dimensional* if there exists a finite subset $\{x_1, \ldots, x_n\} \subset V$ such that $V = \text{span}\{x_1, \ldots, x_n\}$. Otherwise, V is referred to as *infinite-dimensional*. The dimension of the trivial vector space, which consists solely of the zero vector $V = \{0\}$, is defined to be zero.

−   For a finite-dimensional, nonzero vector space V, the *dimension* is determined by the number of vectors in any linearly independent spanning set. Specifically, if V contains an independent set of n vectors but no independent set of $n + 1$ vectors, then the dimension of V is n. An infinite-dimensional vector space is said to have infinite dimension. The dimension of V is denoted by $\dim_{\mathbb{F}} V$ or simply $\dim V$ when the field $\mathbb{F}$ is clear from the context.

−   Assume that $\dim V = n$ and $V = \text{span}\{x_1, \ldots, x_n\}$. The set $\{x_1, \ldots, x_n\}$ is called a *basis* of V if each vector x in V can be uniquely represented as

$$x = \sum_{i=1}^{n} a_i x_i.$$

Hence, for each $x \in V$, there is a unique column vector $a = (a_1, \ldots, a_n)^\top \in \mathbb{F}^n$, where $\mathbb{F}^n$ denotes the vector space of column vectors with n entries in $\mathbb{F}$.

It is often convenient to denote the vector x by the expression $x = [x_1, x_2, \ldots, x_n]a$. An *ordered basis* refers to a basis $\{x_1, \ldots, x_n\}$ along with a specified order, ensuring that x_1 is considered first, and so on.

Direct sum of vector spaces. Let V be a vector space over a field $\mathbb{F}$. Let U and W be subspaces of V. Then V is said to be the *direct sum* of U and W, and we write $V = U \oplus W$, if $V = U + W = \{x + y : x \in U, y \in V\}$ and $U \cap W = \{0\}$. It is proved that $V = U \oplus W$ if and only if for every $v \in V$, there exist unique vectors $x \in U$ and $y \in W$ such that $v = x + y$.

Linear operators. Let U and V be vector spaces over a field $\mathbb{F}$. A map $T : U \to V$ is called a *linear map* or *linear operator* (also known as a *linear transformation*) if for all $a, b \in \mathbb{F}$ and $u, v \in U$, the following property holds:

$$T(au + bv) = aT(u) + bT(v).$$

The set of all linear maps from U to V is denoted by $\mathcal{L}(U, V)$. In the particular case where $U = V$, the notation $\mathcal{L}(V)$ is used.

A linear operator $T : U \to V$ is called a *linear isomorphism* if T is bijective. In this situation, U is said to be *isomorphic* to V, denoted by $U \cong V$. Additionally, the *kernel* and *image* of T, denoted by $\ker(T)$ and $\mathrm{im}(T)$, respectively, are defined as follows:

$$\ker(T) = \{x \in U \mid T(x) = 0\},$$
$$\mathrm{im}(T) = \{T(x) \mid x \in U\}.$$

It is straightforward to show that $\ker(T)$ and $\mathrm{im}(T)$ are subspaces of U and V, respectively. Moreover, the *nullity* and *rank* of T, denoted by $\mathrm{null}(T)$ and $\mathrm{rank}(T)$, are defined as

$$\mathrm{null}(T) = \dim(\ker(T)),$$
$$\mathrm{rank}(T) = \dim(\mathrm{im}(T)).$$

A.9.2 Matrices, notions and notations

A *matrix* is a rectangular array of numbers, symbols, or expressions arranged in rows and columns, typically denoted by $A, B, C, \ldots$. Let S be a set. Throughout, S can be the set $\{0, 1\}$, the set of natural integers $\mathbb{N} := \{1, 2, \ldots\}$, the set of integers $\mathbb{Z} = \{0, \pm 1, \pm 2, \ldots\}$, the set nonnegative integers $\mathbb{Z}_+ := \{0, 1, 2, \ldots\}$, or the sets of real numbers $\mathbb{R}$ and complex numbers $\mathbb{C}$.

Denote by $S^{m \times n}$ the set of $m \times n$ matrices $A = (a_{ij})_{i,j=1}^{m,n}$, where $a_{ij} \in S$:

$$A = \begin{pmatrix} a_{11} & a_{12} & \cdots & a_{1n} \\ a_{21} & a_{22} & \cdots & a_{2n} \\ \vdots & \vdots & \vdots & \vdots \\ a_{m1} & a_{m2} & \cdots & a_{mn} \end{pmatrix}.$$

We list some basic notations and facts about matrices:

- The elements a_{ij} are called the *entries* of A.
- A matrix is called a *zero matrix* if all its entries are zero.
- The $n \times n$ *identity* matrix I_n is defined as $I_n = (\delta_{ij}) \in \{0, 1\}^{n \times n}$, where $\delta_{ij} = 1$ if and only if $i = j$.
- Matrix addition: The sum of matrices

$$A = \begin{pmatrix} a_{11} & a_{12} & \cdots & a_{1n} \\ a_{21} & a_{22} & \cdots & a_{2n} \\ \vdots & \vdots & \ddots & \vdots \\ a_{m1} & a_{m2} & \cdots & a_{mn} \end{pmatrix}$$

and

$$B = \begin{pmatrix} b_{11} & b_{12} & \cdots & b_{1n} \\ b_{21} & b_{22} & \cdots & b_{2n} \\ \vdots & \vdots & \ddots & \vdots \\ b_{m1} & b_{m2} & \cdots & b_{mn} \end{pmatrix}$$

is defined as

$$A + B := \begin{pmatrix} a_{11} + b_{11} & a_{12} + b_{12} & \cdots & a_{1n} + b_{1n} \\ a_{21} + b_{21} & a_{22} + b_{22} & \cdots & a_{2n} + b_{2n} \\ \vdots & \vdots & \ddots & \vdots \\ a_{m1} + b_{m1} & a_{m2} + b_{m2} & \cdots & a_{mn} + b_{mn} \end{pmatrix}.$$

- Matrix multiplication: The product of an $(m \times n)$ matrix

$$A = \begin{pmatrix} a_{11} & a_{12} & \cdots & a_{1n} \\ a_{21} & a_{22} & \cdots & a_{2n} \\ \vdots & \vdots & \ddots & \vdots \\ a_{m1} & a_{m2} & \cdots & a_{mn} \end{pmatrix}$$

and an $(n \times r)$ matrix

$$B = \begin{pmatrix} b_{11} & b_{12} & \cdots & b_{1r} \\ b_{21} & b_{22} & \cdots & b_{2r} \\ \vdots & \vdots & \ddots & \vdots \\ b_{n1} & b_{n2} & \cdots & b_{nr} \end{pmatrix}$$

is defined as the $(m \times r)$ matrix

$$AB = \begin{pmatrix} c_{11} & c_{12} & \cdots & c_{1r} \\ c_{21} & c_{22} & \cdots & c_{2r} \\ \vdots & \vdots & \ddots & \vdots \\ c_{m1} & c_{m2} & \cdots & c_{mr} \end{pmatrix}$$

with the elements c_{ij} computed as

$$c_{ij} = a_{i1}b_{1j} + a_{i2}b_{2j} + \cdots + a_{in}b_{nj}.$$

- Let A be an $m \times n$ matrix, and let B be a $p \times q$ matrix. By the *direct sum* of A and B, written $A \oplus B$, we mean the $(m + p) \times (n + q)$ matrix of the form

$$\begin{pmatrix} A & O \\ O & B \end{pmatrix},$$

where the Os represent zero matrices: O in the top right is an $m \times q$ matrix, whereas O in the bottom left is an $n \times p$ matrix.
- A is called a *square matrix of order n* if it has an equal number of rows and columns (equal to n).
- The entries $a_{11}, a_{22}, \ldots, a_{nn}$ of an $n \times n$ matrix $A = (a_{ij})$ are called the *diagonal entries*.
- A square matrix $A = [a_{ij}]$ is called:
  1. *upper triangular* if $a_{ij} = 0$ for all $i > j$; in other words, all elements below the main diagonal are zero.
  2. *lower triangular* if $a_{ij} = 0$ for all $i < j$; in other words, all elements above the main diagonal are zero.
  3. *strictly upper triangular* if $a_{ij} = 0$ for all $i \le j$; in other words, all elements on and below the main diagonal are zero.
  4. *strictly lower triangular* if $a_{ij} = 0$ for all $i \ge j$. In other words, all elements on and above the main diagonal are zero.
- The *transpose* of A, denoted by $A^\top$, is an $n \times m$ matrix with $A^\top = (a_{ji})_{j,i=1}^{n,m}$.
- Clearly, $(A^\top)^\top = A$.
- A square matrix is called *symmetric* if $A^\top = A$.
- If $A = (a_{ij}) \in S^{n \times n}$, then the entries $a_{11}, a_{22}, \ldots, a_{nn}$ are called the *main diagonal* entries (or, simply, *diagonal* entries).
- If $A = (a_{ij}) \in S^{n \times n}$, then the *trace* of A, denoted by $\operatorname{tr} A$, is the sum of all diagonal elements of A: $\operatorname{tr} A := \sum_{i=1}^{n} a_{ii}$.

- For any $A \in S^{n \times n}$ and a positive integer $k > 1$, $A^k = A \cdot \ldots \cdot A$, where the (matrix) product is taken k times, inductively, that is, $A^k = A \cdot A^{k-1}$.
- We make the convention that $A^0 := I_n$, the $n \times n$ *identity* matrix, and $A^1 = A$.
- A square matrix A is called *nilpotent* if $A^k = 0$ for some $k \in \mathbb{N}$.
- Let $A = (a_{ij})$, $B = (b_{ij}) \in \mathbb{R}^{m \times n}$. Then we say that B *is greater than or equal to* A, denoted by $B \geq A$, if $b_{ij} \geq a_{ij}$ for $i = 1, \ldots, m, j = 1, \ldots, n$.
- If A is symmetric, then A^k is also symmetric.
- A square matrix $P \in \{0, 1\}^{n \times n}$ is called a *permutation* matrix if each row and column of P has exactly one entry equal to 1 and all other entries are equal to 0.
- An $n \times n$ matrix A is called *invertible* if there exists an $n \times n$ matrix B such that $AB = BA = I_n$.
- A matrix $A \in \mathbb{C}^{n \times n}$ is called *orthogonal* if $AA^T = I_n$. In particular, an orthogonal matrix A is always invertible, and $A^{-1} = A^T$.
- Note that if A is a permutation matrix, then we can easily verify that $AA^T = A^T A = I_n$. Therefore A is an orthogonal matrix.
- The set of all permutation matrices $P \in \mathbb{C}^{n \times n}$ is denoted by $\mathcal{P}_n$.
- A matrix $A \in \mathbb{C}^{n \times n}$ is called a *diagonal* matrix if it is a square matrix, whose all off-diagonal entries are 0:

$$\operatorname{diag}(d_1, d_2, \ldots, d_n) = \begin{pmatrix} d_1 & 0 & \ldots & 0 & 0 \\ 0 & d_2 & \ldots & 0 & 0 \\ \vdots & \vdots & \vdots & \vdots & \vdots \\ 0 & 0 & \ldots & 0 & d_n \end{pmatrix}.$$

Example: $\operatorname{diag}(3, -2, 7) = \begin{pmatrix} 3 & 0 & 0 \\ 0 & -2 & 0 \\ 0 & 0 & 7 \end{pmatrix}$.
- Note that the identity matrix is a diagonal matrix with all diagonal entries equal to 1.
- Defining diagonal matrices in a more general case, an $m \times n$ matrix $A = (a_{ij})_{i,j=1}^{m,n}$ is called a diagonal matrix and is denoted by $\operatorname{diag}(d_1, \ldots, d_{\min(m,n)})$ if $a_{ij} = 0$ for $i \neq j$ and $a_{ii} = d_i$ for $i = 1, \ldots, \min(m, n)$.
- Two $n \times n$ matrices A and B are called *similar* if there exists an invertible matrix M such that $B = M^{-1}AM$.
- A square matrix A is called *diagonalizable* if it is similar to a diagonal matrix. This means there exists an invertible matrix P such that

$$P^{-1}AP = D,$$

where D is a diagonal matrix.
- Two $n \times n$ nonnegative matrices A and B are called *diagonally similar* if $B = D^{-1}AD$ for some diagonal matrix D with positive diagonal entries.
- A matrix $P = (p_{ij}) \in \mathbb{R}^{m \times n}$ is called *nonnegative (positive)* and is written $P \geq 0$ ($P > 0$) if $p_{ij} \geq 0$ ($p_{ij} > 0$) for all $i \in [m]$ and $j \in [n]$.

Matrices versus vector spaces. A correspondence between linear operators and matrices is fundamental in linear algebra, as it allows us to study linear operators using the more familiar language of matrices. Consider a field $\mathbb{F}$ and $n, m \in \mathbb{N}$. Then $\mathbb{F}^n$ has a standard basis consisting of the vectors $e_1 = (1, 0, \ldots, 0), e_2 = (0, 1, \ldots, 0), \ldots, e_n = (0, 0, \ldots, 1)$. A linear operator T from $\mathbb{F}^n$ to $\mathbb{F}^m$ can be uniquely represented by an $m \times n$ matrix, known as the *matrix representation* of T. The entries of this matrix are determined by the images of the standard basis vectors in $\mathbb{F}^n$ under T. So there is a one-to-one correspondence between linear operators $T: \mathbb{F}^n \to \mathbb{F}^m$ and $m \times n$ matrices, i. e., $\mathbb{F}^{m \times n}$. Having said this, we can also define parameters related to linear operators, such as rank and nullity, for matrices in a similar manner.

A.9.3 Eigenvectors and eigenvalues of square complex matrices

Determinants. Let S_n denote the group of permutations of the set $[n] := \{1, \ldots, n\}$. For a matrix $A \in \mathbb{C}^{n \times n}$ with entries $A = [a_{ij}]$, the *determinant* of A is defined by the formula

$$\det(A) = \sum_{\omega \in S_n} \text{sign}(\omega)\, a_{1\omega(1)} a_{2\omega(2)} \cdots a_{n\omega(n)}.$$

The determinant function satisfies the following properties for all $A, B \in \mathbb{C}^{n \times n}$ and $a \in \mathbb{C}$:
- $\det(aA) = a^n \det(A)$,
- $\det(AB) = \det(A)\det(B)$,
- $\det(I_n) = 1$.

The following observations about the determinant are also worth noting:
- The determinant of a real (or integer) matrix is a real (or integer) number.
- A significant advantage of the determinant $\det(A)$ is that it can be computed in polynomial time, requiring at most $\frac{2n^3}{3}$ operations.

Invertible matrices. A matrix $A \in \mathbb{C}^{n \times n}$ is called *invertible* if there exists a matrix $B \in \mathbb{C}^{n \times n}$ such that $AB = BA = I_n$. Note that such a matrix B is unique. If $AC = CA = I_n$, then $B = BI_n = B(AC) = (BA)C = I_nC = C$. We denote this matrix B, the inverse of A, by A^{-1}.

In our definition of an invertible matrix A, we required A to be square and the inverse matrix B to satisfy both $AB = I_n$ and $BA = I_n$. In fact, given that A and B are square, either of the conditions $AB = I_n$ or $BA = I_n$ implies the other. If A is invertible, then its *inverse* is denoted by A^{-1}. In this case, A clearly satisfies $AA^{-1} = A^{-1}A = I_n$.

Assume that $A \in \mathbb{C}^{n \times n}$. The matrix A is invertible if and only if $\det A \neq 0$. Furthermore, $\det A = 0$ if and only if the rows (or columns) of A are linearly dependent. Equivalently, $\det A = 0$ if and only if the system $Ax = \mathbf{0}$ has a nontrivial solution $\mathbf{x} \neq \mathbf{0}$.

A nonzero vector $\mathbf{x} \in \mathbb{C}^n$ is called the *right (left) eigenvector* with the corresponding *eigenvalue* λ if $A\mathbf{x} = \lambda\mathbf{x}$ (or $\mathbf{x}^T A = \lambda\mathbf{x}^T$).

The eigenvalues of A are the complex roots of the *characteristic polynomial*

$$\det(zI_n - A) = z^n - (\text{tr } A)z^{n-1} + \cdots + (-1)^n \det A = \prod_{i=1}^{n}(z - \lambda_i).$$

Hence

$$\text{tr } A = \sum_{i=1}^{n} \lambda_i, \quad \det A = \lambda_1\lambda_2\cdots\lambda_n.$$

The set of all distinct eigenvalues of A, denoted by $\text{spec}(A) \subset \mathbb{C}$, is called the *spectrum* of A. For example, $\text{spec}(I_n) = \{1\}$ because $\det(zI_n - I_n) = (z-1)^n$, making 1 the unique root of the characteristic polynomial with multiplicity n. Thus I_n has n eigenvalues, all equal to 1, i. e., $\lambda_1 = \cdots = \lambda_n = 1$.

It is customary to arrange the n eigenvalues of A in one of the following two orders:

$$\mathbb{R}(\lambda_1) \geq \mathbb{R}(\lambda_2) \geq \cdots \geq \mathbb{R}(\lambda_n),$$
$$|\lambda_1| \geq |\lambda_2| \geq \cdots \geq |\lambda_n|. \tag{A.9.1}$$

In the first type of ordering, $\mathbb{R}(\lambda_i)$ denotes the real part of λ_i. The *spectral radius* of A, denoted by $\rho(A)$, is defined as the maximum modulus of all the eigenvalues of A. If the eigenvalues of A are arranged according to the second order above, then $\rho(A) = |\lambda_1|$.

One issue with the eigenvalues of a real matrix A is that some or all of the eigenvalues may be complex (not real). For example, the matrix $A = \left(\begin{smallmatrix} 0 & 1 \\ -1 & 0 \end{smallmatrix}\right)$ has two complex eigenvalues, $\lambda_1 = \sqrt{-1}$ and $\lambda_2 = -\sqrt{-1}$. In this case, the corresponding eigenvectors cannot be chosen to be real.

It is convenient to define

$$\text{Eig}(z, A) := \{\mathbf{x} \in \mathbb{C}^n : A\mathbf{x} = z\mathbf{x}\} \quad \text{for } z \in \mathbb{C},$$

where $\text{Eig}(z, A)$ is defined as the *eigenspace* of A associated with the vector z. Clearly $\text{Eig}(z, A)$ is a subspace of A. Furthermore, $z \in \text{spec } A$ if and only if $\dim \text{Eig}(z, A) > 0$. If z is not an eigenvalue of A, then $\text{Eig}(z, A) = \{\mathbf{0}\}$ is the trivial subspace. For $\lambda \in \text{spec } A$, $\text{Eig}(\lambda, A)$ is the *eigenspace* of A corresponding to λ. Thus any $\mathbf{0} \neq \mathbf{x} \in \text{Eig}(\lambda, A)$ is an eigenvector of A corresponding to λ. The dimension of the vector subspace $\text{Eig}(\lambda, A)$, denoted by $\dim \text{Eig}(\lambda, A)$, is called the *geometric multiplicity* of λ. The multiplicity $m(\lambda)$ of the root λ in the characteristic polynomial $\det(zI_n - A)$ is called the *algebraic multiplicity* of λ. It is known that $\dim \text{Eig}(\lambda, A) \leq m(\lambda)$. λ is called *geometrically simple* if $\dim \text{Eig}(\lambda, A) = m(\lambda)$. Otherwise, λ is called a *defective* eigenvalue.

Assume that $\lambda_i \neq \lambda_j$ for $1 \leq i < j \leq m \leq n$ are m distinct eigenvalues of A. Let $\mathbf{x}_1, \ldots, \mathbf{x}_m \in \mathbb{C}^{n \times n}$ be the corresponding eigenvectors of A: $A\mathbf{x}_i = \lambda_i\mathbf{x}_i$, $i = 1, \ldots, m$. Then it is known that $\mathbf{x}_1, \ldots, \mathbf{x}_m$ are linearly independent. This is equivalent to the statement that the dimension of the subspace $\mathbf{V}$ spanned by all eigenvectors vectors in of A is equal to the sum of the dimensions of all different eigenspaces of A. In another notation, $V = \oplus_{\lambda \in \text{spec } A} \text{Eig}(\lambda, A)$.

Suppose that $m = n$, meaning that the characteristic polynomial of A has n distinct eigenvalues. This situation is typical because if the n^2 entries of A are chosen randomly, then A will have n distinct eigenvalues with probability 1. This condition is equivalent to assuming that each eigenvalue of A is algebraically simple.

Under this assumption, the square matrix $X := (\mathbf{x}_1, \mathbf{x}_2, \ldots, \mathbf{x}_n) \in \mathbb{C}^{n \times n}$ whose ith column is $\mathbf{x}_i$ has a nonzero determinant. Therefore X^{-1} exists. This leads to the following equivalence:

$$A\mathbf{x}_i = \lambda_i \mathbf{x}_i, \; i = 1, \ldots, n \iff AX = X\Lambda \iff A = X\Lambda X^{-1}, \quad \Lambda := \text{diag}(\lambda_1, \ldots, \lambda_n).$$

Such a matrix A is called *diagonalizable*.

Note that A is diagonalizable if and only if for any eigenvalue λ of multiplicity $m \geq 1$ in the characteristic polynomial of A, we have $\text{rank}(\lambda I_n - A) = n - m$. If $A = X\Lambda X^{-1}$, then for any $k \in \mathbb{N}$, $A^k = X\Lambda^k X^{-1}$, making it straightforward to understand the behavior of A^k for $k = 1, 2, \ldots$.

Let $G(A) = (V, E)$ be the digraph induced by A. Let $V = \cup_{i=1}^k V_i$ be a decomposition of V to a union of nonempty disjoint sets such that each $G(V_i)$ is a connected component of G (this does not mean that $G(V_i)$ is strongly connected), that is, for $i \neq j$, there are no directed edges between V_i and V_j, and each undirected graph induced by $G(V_i)$ is connected. Then there exists a permutation matrix $Q \in \{0, 1\}^{n \times n}$ such that

$$QAQ^T = \text{diag}(A_1, \ldots, A_k), \quad A_i \in \mathbb{C}^{n_i \times n_i}, \quad i = 1, \ldots, k. \tag{A.9.2}$$

For simplicity of notation, we assume that we renamed the vertices $[n]$ so that $Q = I_n$, i. e., A is a block diagonal:

$$A = \text{diag}(A_1, \ldots, A_k) = \oplus_{i=1}^k A_i, \quad A_i \in \mathbb{C}^{n_i \times n_i}. \tag{A.9.3}$$

Then $\det(\lambda I_n - A) = \prod_{i=1}^k \det(\lambda I_{n_i} - A_i)$. To find the eigenvalues and eigenvectors of A, it suffices to find the eigenvalues and eigenvectors of each A_i. Specifically, we view $\mathbb{C}^n = \oplus_{i=1}^k \mathbb{C}^{n_i}$ as follows: every vector $\mathbf{w} \in \mathbb{C}^n$ is expressed as $\mathbf{w}_1 \oplus \cdots \oplus \mathbf{w}_k$, where $\mathbf{w}_i \in \mathbb{C}^{n_i}, i = 1, \ldots, k$. In other words, $\mathbf{w}^T = (\mathbf{w}_1^T, \mathbf{w}_2^T, \ldots, \mathbf{w}_k^T)$. (Recall that $\mathbf{w}$ and $\mathbf{w}_1, \ldots, \mathbf{w}_k$ are column vectors!)

We can then describe the eigenspace $\text{Eig}(\lambda, A)$ as follows:

$$\text{Eig}\left(\lambda, \bigoplus_{i=1}^k A_i\right) = \bigoplus_{i=1}^k \text{Eig}(\lambda, A_i) \quad \text{and} \quad \dim \text{Eig}\left(\lambda, \bigoplus_{i=1}^k A_i\right) = \sum_{i=1}^k \dim \text{Eig}(\lambda, A_i) \tag{A.9.4}$$

for all $\lambda \in \text{spec}(A)$. (In fact, the formula holds for all $\lambda \in \mathbb{C}$.)

This means that any eigenvector of A corresponding to the eigenvalue λ is of the form $\mathbf{y}^T = (\mathbf{y}_1^T, \ldots, \mathbf{y}_k^T)$, where the following conditions hold:

- If $\lambda \notin \text{spec}(A_i)$, then $\mathbf{y}_i = 0$.
- If $\lambda \in \text{spec}(A_i)$, then $\mathbf{y}_i$ is either an eigenvector of A_i corresponding to λ, or $\mathbf{y}_i = 0$.
- At least one of the vectors $\mathbf{y}_i$ must be nonzero, and for this $\mathbf{y}_i$, we have $\lambda \in \text{spec}(A_i)$.

Singular values. Let $A \in \mathbb{C}^{m \times n}$. Consider the matrix A^*A. This is a symmetric $n \times n$ matrix, and it is proved that its eigenvalues are real. Furthermore, we may argue that its eigenvalues are nonnegative. Let $\lambda_1, \ldots, \lambda_n$ denote the eigenvalues of $A^T A$, allowing for repetitions, and order these so that $\lambda_1 \geq \lambda_2 \geq \cdots \geq \lambda_n \geq 0$. Define $\sigma_i = \sqrt{\lambda_i}$, so that $\sigma_1 \geq \sigma_2 \geq \cdots \geq \sigma_n \geq 0$. The numbers $\sigma_1 \geq \sigma_2 \geq \cdots \geq \sigma_n \geq 0$ are called the *singular values* of A. We have:
- The number of nonzero singular values of A equals the rank of A.
- In particular, if A is an $m \times n$ matrix with $m < n$, then A has at most m nonzero singular values, as $\text{rank}(A) \leq m$.

A.9.4 Norms

A function $v : \mathbb{C}^n \to \mathbb{R}_+$ is called a *norm* if it satisfies the following properties:
1. *Positivity:* $v(\mathbf{x}) > 0$ for $\mathbf{x} \neq \mathbf{0}$, and $v(\mathbf{0}) = 0$.
2. *Homogeneity:* $v(a\mathbf{x}) = |a|v(\mathbf{x})$ for any scalar $a \in \mathbb{C}$ and $\mathbf{x} \in \mathbb{C}^n$.
3. *Triangle inequality:* $v(\mathbf{x} + \mathbf{y}) \leq v(\mathbf{x}) + v(\mathbf{y})$ for all $\mathbf{x}, \mathbf{y} \in \mathbb{C}^n$.

The most common examples of norms are the l_p *norms* defined as follows:

$$\|\mathbf{x}\|_p = \|\mathbf{x}^T\|_p := \left(\sum_{i=1}^{n} |x_i|^p \right)^{\frac{1}{p}} \quad \text{for } p \in [1, \infty), \tag{A.9.5}$$

where $\mathbf{x} = (x_1, \ldots, x_n)$. Note that sometimes in subsequent discussions, we use l_p norms for row vectors.

The Euclidean norm $\|\mathbf{x}\|_2 = \|\mathbf{x}^*\|_2$ is equivalent to the norm $\|\mathbf{x}\| = \|\mathbf{x}^*\| = \sqrt{\mathbf{x}^*\mathbf{x}}$ defined in Section 5.5.

Consider $A \in \mathbb{C}^{m \times n}$ as an operator $A : \mathbb{C}^n \to \mathbb{C}^m$ given by $\mathbf{x} \mapsto A\mathbf{x}$. The *operator l_p norm* of A is defined as

$$\|A\|_p := \max_{\|\mathbf{x}\|_p=1, \mathbf{x} \in \mathbb{C}^n} \|A\mathbf{x}\|_p = \max_{\mathbf{x} \neq 0 \in \mathbb{C}^n} \frac{\|A\mathbf{x}\|_p}{\|\mathbf{x}\|_p} \quad \text{for } A \in \mathbb{C}^{m \times n} \text{ and } p \in [1, \infty]. \tag{A.9.6}$$

The norm $\| \cdot \|_p$ on $\mathbb{C}^{m \times n}$ satisfies $\|I_n\|_p = 1$ for all $n \in \mathbb{N}$.

From the definition of the operator norm it follows that

$$\|A\mathbf{x}\|_p \leq \|A\|_p \|\mathbf{x}\|_p \quad \text{for all } \mathbf{x} \in \mathbb{C}^n, A \in \mathbb{C}^{m \times n},$$
$$\|A + B\|_p \leq \|A\|_p + \|B\|_p \quad \text{for all } A, B \in \mathbb{C}^{m \times n},$$

$$\|AB\|_p \le \|A\|_p\|B\|_p \quad \text{for all } A \in \mathbb{C}^{m\times n},\ B \in \mathbb{C}^{n\times l}, \tag{A.9.7}$$

$$\left\|\text{diag}(d_1,\dots,d_{\min(m,n)})\right\|_p = \max_{i\in[1,\min(m,n)]} |d_i| \quad \text{for } \text{diag}(d_1,\dots,d_{\min(m,n)}) \in \mathbb{C}^{m\times n},$$

$$\|A^{l+q}\|_p \le \|A^l\|_p\|A^q\|_p \quad \text{for all } l,q \in \mathbb{Z}_+ \text{ and } A \in \mathbb{C}^{n\times n}.$$

We also define the *infinity norm* as follows. Let $\mathbf{x} \in \mathbb{C}^n$. Then

$$\|\mathbf{x}\|_\infty = \|\mathbf{x}^\top\|_\infty := \max_{1\le i\le n} |x_i|.$$

Moreover, for $A \in \mathbb{C}^{m\times n}$, we define the *subordinate matrix infinity norm* as

$$\|A\|_\infty = \max\left\{\frac{\|A\mathbf{x}\|_\infty}{\|\mathbf{x}\|_\infty} : \mathbf{x} \in \mathbb{C}^n, \mathbf{x} \ne \mathbf{0}\right\}.$$

A.9.5 Inner product spaces

Let V be a vector space over a field F. The function $\langle\cdot,\cdot\rangle : V \times V \to F$ is referred to as an *inner product* if it satisfies the following conditions for all $x,y,z \in V$ and $a \in F$:
1. *Conjugate symmetry:*

$$\langle x,y\rangle = \overline{\langle y,x\rangle}.$$

2. *Linearity in the first argument:*

$$\langle ax + y, z\rangle = a\langle x,z\rangle + \langle y,z\rangle.$$

3. *Positive-definiteness:*

$$\langle x,x\rangle \ge 0; \quad \langle x,x\rangle = 0 \quad \text{if and only if} \quad x = 0.$$

The vector space V equipped with the inner product $\langle\cdot,\cdot\rangle$ is known as an *inner product space*, abbreviated as **IPS**.

A.9.6 Tensor product of vector spaces (matrices)

Let V, W, and X be vector spaces over a field $\mathbb{F}$. A *bilinear map* from $V \times W$ to X is a function $T : V \times W \to X$ that satisfies the following properties:
1. For all $v_1, v_2 \in V$, $w \in W$, and $a \in \mathbb{F}$, we have

$$T(av_1 + v_2, w) = aT(v_1, w) + T(v_2, w).$$

2. For all $v \in V, w_1, w_2 \in W$, and $a \in \mathbb{F}$, we have

$$T(v, aw_1 + w_2) = aT(v, w_1) + T(v, w_2).$$

A *tensor product* of V and W is a vector space $V \otimes W$ equipped with a bilinear map $\varphi : V \times W \to V \otimes W$. This tensor product satisfies the property that for every vector space X and every bilinear map $T : V \times W \to X$, there exists a unique linear map $f : V \otimes W \to X$ such that

$$T = f \circ \varphi.$$

In other words, specifying a linear map from $V \otimes W$ to X is equivalent to specifying a bilinear map from $V \times W$ to X.

When $V \otimes W$ is given as a tensor product, we use the notation $v \otimes w$ to denote $\varphi(v, w)$. It is essential to recognize that a tensor product consists of two key counterparts, the vector space $V \otimes W$ and the bilinear map $\varphi : V \times W \to V \otimes W$.

For a proof of the existence of tensor products, we refer to [16], where an explicit tensor product is constructed. The following theorem summarizes the fundamental properties of tensor products.

Theorem A.19. *Let V and W be vector spaces over a field $\mathbb{F}$.*
1. *Any two tensor products of V and W are isomorphic.*
2. *The vector spaces V and W always have a tensor product.*
3. *If $\{v_1, \ldots, v_n\}$ is a basis for V and $\{w_1, \ldots, w_m\}$ is a basis for W, then $\{v_i \otimes w_j\}_{1 \le i \le n, 1 \le j \le m}$ forms a basis for $V \otimes W$. Consequently, elements of $V \otimes W$ can be represented as $n \times m$ matrices with entries in $\mathbb{F}$.*

A.9.7 The Gram–Schmidt process

The *Gram–Schmidt process* (GSP) is a method for orthonormalizing a set of vectors in an inner product space. Let V be an **IPS**, and let $S = \{\mathbf{x}_1, \ldots, \mathbf{x}_m\} \subset V$ be a finite (possibly empty) set. The set $\tilde{S} = \{\mathbf{e}_1, \ldots, \mathbf{e}_p\}$ is the orthonormal set (with $p \ge 1$) or the empty set (with $p = 0$) obtained from S using the following recursive steps:
1. If $\mathbf{x}_1 = 0$, then remove it from S. Otherwise, replace $\mathbf{x}_1$ with $\frac{\mathbf{x}_1}{\|\mathbf{x}_1\|}$.
2. Assume that $\{\mathbf{x}_1, \ldots, \mathbf{x}_k\}$ is an orthonormal set and $1 \le k < m$. Let

$$\mathbf{y}_{k+1} = \mathbf{x}_{k+1} - \sum_{i=1}^{k} \langle \mathbf{x}_{k+1}, \mathbf{x}_i \rangle \mathbf{x}_i.$$

If $\mathbf{y}_{k+1} = 0$, then remove $\mathbf{x}_{k+1}$ from S. Otherwise, replace $\mathbf{x}_{k+1}$ with $\frac{\mathbf{y}_{k+1}}{\|\mathbf{y}_{k+1}\|}$.

> **Remark A.20.** Let V be an IPS, and let $S = \{\mathbf{x}_1, \ldots, \mathbf{x}_n\} \subset V$ be a set of n linearly independent vectors. Then, the Gram–Schmidt algorithm on S is given as follows:
>
> $$\mathbf{y}_1 := \mathbf{x}_1, \quad r_{11} := \|\mathbf{y}_1\|, \quad \mathbf{e}_1 := \frac{\mathbf{y}_1}{r_{11}},$$
>
> $$r_{ji} := \langle \mathbf{x}_i, \mathbf{e}_j \rangle, \quad j = 1, \ldots, i-1,$$
>
> $$\mathbf{p}_{i-1} := \sum_{j=1}^{i-1} r_{ji}\mathbf{e}_j, \quad \mathbf{y}_i := \mathbf{x}_i - \mathbf{p}_{i-1},$$
>
> $$r_{ii} := \|\mathbf{y}_i\|, \quad \mathbf{e}_i := \frac{\mathbf{y}_i}{r_{ii}}, \quad i = 2, \ldots, n.$$
>
> In particular, $\mathbf{e}_i \in S_i$, and $\|\mathbf{y}_i\| = \mathrm{dist}(\mathbf{x}_i, S_{i-1})$, where $S_i = \mathrm{span}\{\mathbf{x}_1, \ldots, \mathbf{x}_i\}$, for $i = 1, \ldots, n$, and $S_0 = \{0\}$.

> **Remark A.21.** Any (ordered) basis in a finite-dimensional inner product space V induces an orthonormal basis via the Gram–Schmidt algorithm.

A.10 Basic topology

We briefly review fundamental concepts of basic topology. For a detailed treatment, the reader is referred to [33].

A *topological space* is defined as an ordered pair $(X, \mathcal{I})$, where X is a set, and $\mathcal{I}$ is a collection of subsets of X that satisfies the following properties:
1. $\emptyset$ and X are elements of $\mathcal{I}$.
2. If $U, V \in \mathcal{I}$, then $U \cap V \in \mathcal{I}$.
3. If $\{U_\alpha \mid \alpha \in I\} \subset \mathcal{I}$, then $\bigcup_{\alpha \in I} U_\alpha \in \mathcal{I}$.

The collection $\mathcal{I}$ is called a *topology* on X, and the pair $(X, \mathcal{I})$ is referred to as a topological space. The elements of $\mathcal{I}$ are known as *open* sets. A subset $F \subset X$ is called *closed* if its complement $X \setminus F$ is open. Although the official notation for a topological space includes the topology $\mathcal{I}$, this notation is often omitted when the topology is clear from the context.

The following observations follow immediately from the definition of a topological space:
- A straightforward induction shows that every finite intersection $U_1 \cap \cdots \cap U_k$ of open sets is also open.
- An arbitrary (infinite) union of open sets is not necessarily open, and determining whether it is can often be challenging.
- Being open and closed are not mutually exclusive properties. In fact, some subsets can be both open and closed.
- The collection of closed subsets in a topological space uniquely determines the topology, just as the collection of open sets does. Therefore, to define a topology on a set, it suffices to specify a collection of subsets that satisfy the properties outlined below as fundamental properties of closed sets:

- Ø and X are closed sets.
- If $F, G \subset X$ are closed, then $F \cup G$ is also closed.
- If $\{F_\alpha \mid \alpha \in I\}$ is a collection of closed subsets, then $\bigcap_{\alpha \in I} F_\alpha$ is closed as well.

- A *base* of a topology is a collection $\mathcal{B}$ of sets in a topological space $(X, \mathcal{I})$ that generates the topology $\mathcal{I}$. For any base $\mathcal{B}$, the union of the sets in $\mathcal{B}$ is equal to X. In other words, for any element $x \in X$, there exists a basis set $A \in \mathcal{B}$ such that $x \in A$. Moreover, for any two sets $A, B \in \mathcal{B}$, given an element $x \in A \cap B$, there exists another set $C \in \mathcal{B}$ such that $x \in C \subset A \cap B$.

Euclidean topology. Let $X = \mathbb{R}^n = \{(x_1, \ldots, x_n) \mid x_i \in \mathbb{R}\}$. As mentioned in Appendix A.7, we define the open ball with center $x \in \mathbb{R}^n$ and radius $\epsilon > 0$ as

$$B(x, \epsilon) \overset{\text{def}}{=} \{y \in \mathbb{R}^n \mid \|x - y\|_2 < \epsilon\},$$

where

$$\|x\|_2 \overset{\text{def}}{=} \sqrt{\sum_{i=1}^{n} x_i^2}.$$

The *Euclidean* (or *classical) topology* on $\mathbb{R}^n$ is given by the collection of arbitrary unions of open balls. More formally, a subset $U \subset \mathbb{R}^n$ is open in the Euclidean topology if and only if there exists a collection of open balls $\{B(x_\alpha, \epsilon_\alpha) \mid \alpha \in I\}$ such that

$$U = \bigcup_{\alpha \in I} B(x_\alpha, \epsilon_\alpha).$$

Throughout, $\mathbb{R}^n$ is always considered with the classical topology unless otherwise mentioned.

Standard Tychonoff product topology. Let $\{X_\alpha\}_{\alpha \in \Lambda}$ be a collection of sets indexed by the set Λ. The Cartesian product $\prod_{\alpha \in \Lambda} X_\alpha$ is the collection of mappings from the index set Λ to the union $\bigcup_{\alpha \in \Lambda} X_\alpha$ such that each index $\alpha \in \Lambda$ is mapped to a member of X_α. For $x \in \prod_{\alpha \in \Lambda} X_\alpha$, denote $x(\alpha)$ as x_α, the αth component of x. For each $\alpha_0 \in \Lambda$, define the α_0-projection mapping $\pi_{\alpha_0} : \prod_{\alpha \in \Lambda} X_\alpha \to X_{\alpha_0}$ as $\pi_{\alpha_0}(x) = x_{\alpha_0}$ for $x \in \prod_{\alpha \in \Lambda} X_\alpha$. Now let $\{(X_\alpha, \mathcal{I}_\alpha)\}_{\alpha \in \Lambda}$ be a collection of topological spaces indexed by an arbitrary set Λ. The *Tychonoff product topology* on the Cartesian product $\prod_{\alpha \in \Lambda} X_\alpha$ is the topology that has as a basis set of the form $\prod_{\alpha \in \Lambda} O_\alpha$, where each $O_\alpha \in \mathcal{I}_\alpha$ is open, and $O_\alpha = X_\alpha$ except for finitely many α.

Compactness. The concept of a compact space significantly generalizes the idea of closed and bounded sets in Euclidean spaces. For a detailed discussion on compactness in metric spaces and its relationship with bounded and closed sets, see Appendix A.7.1. We begin with a few introductory definitions:

- A *cover* or *covering* $\mathcal{C}$ of a topological space X is a collection of subsets of X whose union is X.

- A covering C is called *open* if each element of C is an open subset of X.
- A *subcover* of a covering C is a subset of C whose union still covers X.
- Let $A \subset X$ be an arbitrary subspace. A *cover* of A is a collection of subsets of X whose union contains A.
- A topological space X is said to be *compact* if every open cover of X has a finite subcover. This is known as the *Heine–Borel property*.
- Let C be an arbitrary collection of subsets of X. We say that C has the *finite intersection property* if the intersection of every finite subcollection of C is nonempty.
- It is proved that a topological space X is compact if and only if for every collection C of closed subsets of X with the finite intersection property, the intersection of all sets in C is nonempty.
- If $X_1, \ldots, X_k$ are compact topological spaces, then it is proved that $X_1 \times \cdots \times X_k$ is compact as well. A remarkable and nontrivial result, known as *Tychonoff's theorem*, states that an arbitrary product of compact topological spaces is compact. The proof of this theorem requires advanced tools from set theory.

List of symbols

- $\mathbb{N}$, set of natural numbers
- $\mathbb{Z}$, ring of integers
- $\mathbb{Q}$, field of rational numbers
- $\mathbb{R}$, field of real numbers
- $\mathbb{C}$, field of complex numbers
- $\mathbb{Z}_+$, all nonnegative integers
- $\mathbb{R}_+$, all nonnegative real numbers
- $z = x + y\mathbf{i}$, complex number
- $\bar{z} := x - y\mathbf{i}$, conjugate of $z = x + y\mathbf{i}$
- $p \Rightarrow q$, p implies q
- $p \Leftrightarrow q$, p if and only if q
- $\approx$, approximately equal to
- $\in$, is an element of
- $\notin$, is not an element of
- $\subset$, is a subset of
- $\varnothing$, empty set
- $A \cup B$, union of A and B
- $A \cap B$, intersection of A and B
- $A \backslash B$, A minus B (set difference)
- A^c, complement of A
- $\bigcup_{i=1}^{n} A_i$, union of A_i from $i = 1$ to n
- $\bigcap_{i=1}^{n} A_i$, intersection of A_i from $i = 1$ to n
- $A \times B$, Cartesian product of A and B
- $|A|$, cardinality of A, order of A
- 2^A or $P(A)$, set of all subsets of A
- $f : A \rightarrow B$, f is a function from A to B
- $f(A)$, image of A under f
- $f^{-1}(A)$, preimage of A under f
- $x \sim y$ or $(x, y) \in R$, x is in relation with y
- $A \sim B$, sets A and B are equinumerous
- $\sum_{i=1}^{n} i$, sum of i from i equals 1 to n
- $\prod_{i=1}^{n} i$, product of i from i equals 1 to n
- $[n] := \{1, \ldots, n\}$
- $n! = 1 \times 2 \times \cdots \times n$, factorial of n
- $\binom{n}{k} = \frac{n!}{k!(n-k)!}$, n choose k
- $\gcd(a, b)$, greatest common divisor of a and b
- $\lceil x \rceil$, ceiling function of x
- $\lfloor x \rfloor$, floor function of x
- $\mathbf{Pr}(A)$, probability of A
- $\mathbf{Pr}(X = a, Y = b)$, joint probability of X and Y
- $(\Omega, \mathcal{F}, \mathbf{Pr})$, probability space
- $\mathbf{Pr}(A|B)$, conditional probability
- $\mathbf{Pr}(X_m \mid X_0, X_1, \ldots, X_{m-1})$, probability distribution of X_m given the values of $X_0, \ldots, X_{m-1}$.
- w.h.p, with high probability
- $\mu^{(m)} := (\mu_1^{(m)}, \ldots, \mu_n^{(m)})$, row probability vector
- $X : \Omega \rightarrow \mathbb{R}$, X is a random variable.
- R_X, range of a random variable X

https://doi.org/10.1515/9783111337388-009

- CDF, cumulative distribution function
- PMF, probability mass function
- $E(X)$, expected value of X
- $E(X^k)$, k-moment of X
- $Var(X)$, variance of X
- $Cov(X, Y)$, covariance of X and Y
- σ_X, standard deviation of X
- $X \sim Ber(p)$, X is a Bernoulli random variable
- $X \sim Bin(n, p)$, X is a binomial random variable
- $X \sim Pu(a)$, X is a Poisson random variable
- i. i. d, independent identically distributed random variables
- $\lim_{t \to a} f(t)$, limit of f as t approaches a
- $\lim_{t \searrow a} f(t)$, limit of f as t approaches a from right
- $\lim_{t \nearrow a} f(t)$, limit of f as t approaches a from left
- $r \leqslant p$, r is much less than p
- $r \approx p$, r is almost equal to p
- (X, d), metric space
- $B_d(x, \epsilon)$, ball of radius ϵ centered at x
- $\partial(A)$, set of all boundary points of A
- A', accumulation points or limit points of A
- $\bar{A}$, closure of A
- $diam(A)$, diameter of A
- $\frac{\partial f}{\partial x}$, partial derivative of f with respect to x
- $C^k(U)$, space of k times continuously differentiable functions on U
- $C^\infty(U)$, space of smooth functions on U
- $(X, \mathcal{I})$, topology $\mathcal{I}$ on a set X
- $G = (V, E)$, graph G with vertex set V and edge set E
- $G \cong H$, a graph G is isomorphic to a graph H
- $G[Y]$, subgraph of G induced by a vertex set Y
- $G[S]$, subgraph of G induced by the edges in a set S
- $G(A)$, graph induced by a matrix A
- $A(G)$, adjacency matrix of A
- $G - e$, edge deletion in a graph G
- $G + e$, edge addition in a graph G
- $\bigcup_{i=1}^{k} G_i$, disjoint union of graphs $G_1, \ldots, G_k$
- K_n, complete graph in n vertices
- G^c, complement of a graph G
- $\Gamma(v)$, neighborhood of a vertex v
- $deg(v)$, degree of a vertex v
- $\mathcal{W}_p(i, j)$, set of all walks on G from i to j in p steps
- $u \sim v$, vertex v is connected to a vertex u
- $dist(u, v)$, distance of vertices u and v
- g_G, girth of a graph G
- $D(G)$, diameter of a graph G
- $\mathcal{X}(G)$, chromatic number of a graph G
- $deg_{out}(v)$, out degree of a vertex v
- $deg_{in}(v)$, in degree of a vertex v
- $G = (V, E_{undir})$, undirected graph G
- $G = (V, E_{max})$, maximal directed graph G

- $G = (V, E_{\text{orient}})$, minimal directed graph G
- $G_{\text{rdc}} = (V_{\text{rdc}}, E_{\text{rdc}})$, reduced graph
- $\text{Cay}(G, X)$, Cayley graph
- $\mathcal{G}_{n,p}$, Gilbert's random graph model
- $\mathcal{G}_{n,m}$, Erdös–Rényi's random graph model
- $\text{supp } a$, support of a distribution a
- Π_n, set of all distributions $a = (a_1, \ldots, a_n)$ on $[n]$.
- $\text{conv } (\mathbf{x}_1, \ldots, \mathbf{x}_p)$, convex hull spanned by $\mathbf{x}_1, \ldots, \mathbf{x}_p$
- $A = (a_{ij})_{i,j=1}^{m,n}$, m by n matrix with entries a_{ij}
- $\mathbb{R}_+^{n \times n}$, set of $n \times n$ matrices with nonnegative entries
- $\mathbb{R}^{n \times n}$, set of $n \times n$ matrices with real entries
- $\mathbb{C}^{n \times n}$, $n \times n$ matrices with complex entries
- $\mathbb{F}^{n \times m}$, $n \times m$ matrices with entries in the field $\mathbb{F}$
- $A + B$, matrix addition
- AB, matrix multiplication
- A^{-1}, inverse of a matrix A
- I_n, $n \times n$ identity matrix
- $\text{tr } A$, trace of a matrix A
- $\det(A)$, determinant of a matrix A
- $B \geq A$, a matrix B is greater than a matrix A
- $\mathbf{1} = (1, \ldots, 1)^\top$
- $\text{diag}(d_1, d_2, \ldots, d_n)$, diagonal matrix with diagonal entries $d_1, d_2, \ldots, d_n$
- $m(\lambda)$, multiplicity of an eigenvalue λ
- $\text{spec}(A)$, set of all distinct eigenvalues of A
- $\text{Eig}(z, A)$, eigenspace of A associated with a vector z
- $U \oplus V$, direct sum of vector spaces U and V
- $\bigoplus_{i=1}^{n} V_i$, direct sum of vector spaces $V_1, \ldots, V_n$
- $V \otimes W$, tensor product of V and W
- $A^\top$, transpose of a matrix A
- $\bar{A}$, conjugate of a matrix $A \in \mathbb{C}^{n \times n}$
- $A^* = \bar{A}^\top$, conjugate transpose of a matrix A
- $\mathbb{H}_n$, set of all $n \times n$ Hermitian matrices
- $\mathbb{S}_n(\mathbb{R}) = \mathbb{H}_n \cap \mathbb{R}^{n \times n}$, set of real symmetric matrices in $\mathbb{R}^{n \times n}$
- $\mathbb{S}_n(\mathbb{R}_+)$, set symmetric matrices in $\mathbb{R}_+^{n \times n}$
- $\mathbb{U}_n$, set of unitary matrices
- $\mathbb{O}_n := \mathbb{U}_n \cap \mathbb{R}^{n \times n}$, group of real orthogonal matrices
- Δ_n, set of $n \times n$ diagonal matrices with positive diagonal entries
- $\|\mathbf{x}\|$, norm of vector $\mathbf{x}$
- $\|\mathbf{x}\|_p$, l_p norm of a vector $\mathbf{x}$
- $\|\mathbf{x}\|_\infty$, infinity norm of a vector $\mathbf{x}$
- $\|A\|_p$, l_p norm of a matrix A
- $\|A\|_\infty$, infinity norm of a matrix A
- $\langle \cdot, \cdot \rangle$, inner product
- **IPS**, inner product space
- GSP, Gram–Schmidt process
- SVD, singular value decomposition
- $\rho(A)$, spectral radius of a matrix A
- $\mathcal{S}_n$, set of all $n \times n$ stochastic matrics
- $\text{span}\{x_1, \ldots, x_n\}$, subspace spanned by $x_1, \ldots, x_n$.

- $\dim_{\mathbb{F}} V$, dimension of a vector space V over a field $\mathbb{F}$
- $T : U \to V$, linear operator from U to V
- $U \cong V$, vector space U is isomorphic to a vector space V
- $\mathcal{L}(U, V)$, set of all linear operators from U to V
- $\mathcal{L}(V)$, set of all linear operators from V to itself
- $\ker(T)$, kernel of a linear operator T
- $\operatorname{im}(T)$, image of a linear operator T
- $\operatorname{null}(T)$, nullity of a linear operator T
- $\operatorname{rank}(T)$, rank of a linear operator T
- $(G, +)$, group G with operation $+$
- $(G, \cdot)$, group G with operation $\cdot$
- $G \cong H$, group G is isomorphic to a group H
- S_n, group of permutations on $[n]$
- $\operatorname{ord}(a)$, order of element a in group setting
- $|G|$, order of a group G
- $\langle a \rangle$, cyclic group generated by a
- $\langle X \rangle$, group generated by a set X
- $(R, +, \cdot)$, ring R with operations $+$ and $\cdot$
- $(\mathbb{F}, +, \cdot)$, field $\mathbb{F}$ with operations $+$ and $\cdot$
- $l(w)$, length of a word w
- **MFPT**, mean first passage time
- **MRT**, mean recurrence time
- **MCMC**, Markov chain Monte Carlo method
- $\mathbf{a}_m = (a_1, a_2, \ldots, a_m)$, word of length m
- $\phi : V \to \{0, 1\}$, configuration
- I_G, total number of feasible configurations for G
- $n(\Psi)$, number of occupied vertices in configuration Ψ
- $P_{\Psi, \Psi'}$, transition probability from state Ψ to state Ψ'
- $\mathbf{d} = \mathbf{dis}(\Psi, \Psi')$, number of vertices in which Ψ and Ψ' differ
- $h(G)$, Shannon capacity of channel represented by G
- $[n]^{\mathbb{N}}$, set of all mappings $\mathbf{a} : \mathbb{N} \to [n]$
- $[n]^{\mathbb{N}}(G) = \{\mathbf{a} = (a_1, \ldots) \in [n]^{\mathbb{N}} : (a_i, a_{i+1}) \in E, i = 1, 2, \ldots\}$
- $W(q)$, set of all permissible words of length q
- $\mathcal{W}_{\mathrm{per}}(q)$, set of all projections onto the first q coordinates of q-periodic words in $[n]^{\mathbb{N}}(G)$
- $l_q = |W(q)|$
- $l_{q,\mathrm{per}} = |W_{\mathrm{per}}(q)|$
- $H(\pi)$, entropy of distribution π
- $Z(m, \mathbf{u}, G)$, grand partition function
- $P(\mathbf{u}, G)$, pressure function
- $h_G^*(\mathbf{p})$, entropy for the color density $\mathbf{p}$
- $\delta(G, B)$, Hausdorff dimension
- **SSFT**, strict subshift of finite type
- d_c, chordal distance
- $\mathcal{T} = (V, E)$, infinite tree
- $\operatorname{Aut}(\mathcal{T})$, group of isometries of infinite $\mathcal{T}$

Bibliography

[1]   D. Aldous and J. A. Fill, *Reversible Markov Chains and Random Walks on Graphs*, Unfinished monograph, available at https://www.stat.berkeley.edu/users/aldous/RWG/book.html.

[2]   J. L. R. Alfonsin, *The Diophantine Frobenius Problem*, Oxford Lecture Series in Mathematics. Oxford University Press.

[3]   N. Alon and M. Krivelevich, The Concentration of the Chromatic Number of Random Graphs, *Combinatorica* 17 (1997), no. 3, 303–313.

[4]   N. Alon and J. Spencer, *The Probabilistic Method*, Fourth edition. Wiley Series in Discrete Mathematics and Optimization. John Wiley-Sons, Inc., Hoboken, NJ, 2016.

[5]   R. B. Bapat and T. E. S. Raghavan, *Nonnegative Matrices and Applications*, Encyclopedia of Mathematics and its Applications, 64. Cambridge University Press, Cambridge, 1997.

[6]   B. Bollobás, *Graph Theory, An Introductory Course*, Graduate Texts in Mathematics, 63. Springer-Verlag, New York–Berlin, 1979.

[7]   B. Bollobás, *Modern Graph Theory*, Graduate Texts in Mathematics, 184. Springer-Verlag, New York, 1998.

[8]   B. Bollobás, *Random Graphs*, 2nd edition. Cambridge Studies in Advanced Mathematics, No. 73. 2001.

[9]   B. Bollobas and A. Thomason, Threshold Functions, *Combinatorica* 7 (1987), 35–38.

[10] A. Brauer, On a Problem of Partitions, *Amer. J. Math.* 64 (1942), 299–312.

[11] G. M. Engel and H. Schneider, Diagonal Similarity and Equivalence for Matrices over Groups with 0, *Czechoslovak Math. J.* 25 (1975), 389–403.

[12] P. Erdös and A. Rényi, On the Evolution of Random Graphs, *Magyar Tud. Akad. Mat. Kut. Int. Közl.* 5 (1960), 17–61.

[13] S. Friedland, Computing the Hausdorff Dimension of Subshifts Using Matrices. *Linear Algebra Appl.* 273 (1998), 133–167.

[14] Discrete Lyapunov Exponents and Hausdorff Dimension. *Ergodic Theory Dynam. Systems* 20 (2000), no. 1, 145–172.

[15] S. Friedland, *Matrices–Algebra, Analysis, and Applications*, World Scientific Publishing Co., Hackensack, NJ, 2016.

[16] S. Friedland and M. Aliabadi, *Linear Algebra and Matrices*, Society for Industrial and Applied Mathematics (SIAM), Philadelphia, PA, 2018.

[17] S. Friedland and U. N. Peled, The pressure, densities and first-order phase transitions associated with multidimensional SOFT, *Notions of Positivity and the Geometry of Polynomials*, 179–220, Trends Math. Birkhäuser/Springer Basel AG, Basel, 2011.

[18] S. Friedland and S. Karlin, Some Inequalities for the Spectral Radius of Nonnegative Matrices and Applications, *Duke Math. J.* 42 (1975), 459–490.

[19] S. Friedland and H. Schneider, The Growth of Powers of Nonnegative Matrices, *SIAM J. Algebraic Discrete Methods* 1 (1980), 185–200.

[20] E. N. Gilbert, Random Graphs, *Annals of Mathematical Statistics* 30 (1959), 1141–1144.

[21] D. S. Gunderson, *Handbook of Mathematical Induction: Theory and Applications*, CRC Press, Boca Raton, FL, 2011.

[22] O. Häggström, *Finite Markov Chains and Algorithmic Applications*, Cambridge University Press, Cambridge, UK, 2002.

[23] I. N. Herstein, *Abstract Algebra*, Macmillan Publishing Company, New York, 1986.

[24] R. A. Horn and C. R. Johnson, *Matrix Analysis*, Second edition. Cambridge University Press, Cambridge, 2013.

[25] J. G. Kemeny and J. L. Snell, *Finite Markov Chains*, The University Series in Undergraduate Mathematics. D. Van Nostrand Co., Inc., Princeton, N. J.–Toronto–London–New York, 1960.

[26] H. H. Jennings and J. L. Moreno, Statistics of Social Configurations, *Sociometry* 1 (1938), 342–374.

https://doi.org/10.1515/9783111337388-010

[27] J. F. C. Kingman, A Convexity Property of Positive Matrices, *Quart. J. Math. Oxford Ser. (2)* 12 (1961), 283–284.

[28] A. N. Kolmogorov, *Foundations of the Theory of Probability*, Chelsea Publishing Co., New York, 1956.

[29] D. P. Landau and K. Binder, *A Guide to Monte Carlo Simulations in Statistical Physics*, Cambridge University Press, Cambridge, UK, 2000.

[30] R. Lyons and Y. Peres, *Probability on Trees and Networks*, Cambridge University Press, New York, US, 2016.

[31] A. Markov, Recherches sur les Valeurs Extrêmes des Intégrales et sur l'Interpolation, *Acta Math.* 28 (1904), 1–64.

[32] H. Minc, *Nonnegative Matrices*, Wiley-Interscience Series in Discrete Mathematics and Optimization. A Wiley-Interscience Publication. John Wiley and Sons, Inc., New York, 1988.

[33] J. Munkres, *Topology*, Second edition. Prentice Hall, Inc., Upper Saddle River, NJ, 2000.

[34] W. Parry, Intrinsic Markov Chains, *Trans. Amer. Math. Soc.* 112 (1964), 5–65.

[35] J. B. Roberts, Note on Linear Forms, *Proc. Amer. Math. Soc.* 7 (1956), 465–469.

[36] U. G. Rothblum, Algebraic Eigenspaces of Nonnegative Matrices, *Linear Algebra Appl.* 12 (1975), 281–292.

[37] W. Rudin, *Principles of Mathematical Analysis*, 3rd edition. International Series in Pure and Applied Mathematics. McGraw-Hill, New York–Auckland-Düsseldorf, 1976.

[38] E. S. Selmer, On the Linear Diophantine Problem of Frobenius, *J. Reine Angew. Math.* 293/294 (1977), 1–17.

[39] R. Schindler, *Set Theory: Exploring Independence and Truth*, Universitext. Springer, Cham, 2014.

[40] J. Spencer, *Ten Lectures on the Probabilistic Method*, CBMS-NSF Regional Conference Series in Applied Mathematics, 52. Society for Industrial and Applied Mathematics (SIAM), Philadelphia, PA, 1987.

[41] M. Z. Spivey, *The Art of Proving Binomial Identities*, CRC Press, Boca Raton, FL, 2019.

[42] H. M. Stark, *An Introduction to Number Theory*, MIT Press, Cambridge, Mass.–London, 1978.

[43] J. L. Taylor, *Foundations of Analysis*, Pure and Applied Undergraduate Texts, 18. American Mathematical Society, Providence, RI, 2012.

[44] P. Walters, *An Introduction to Ergodic Theory*, Graduate Texts in Mathematics, 79. Springer-Verlag, New York–Berlin, 1982.

[45] D. West, *Introduction to Graph Theory*, Prentice Hall, Upper Saddle River, NJ, 1996.

[46] H. H. Wu and S. Wu, Various Proofs of the Cauchy–Schwarz Inequality, *Octogon Math. Mag.* 17 (2009), 221–229.

About the authors

Shmuel Friedland is a Professor Emeritus at the University of Illinois at Chicago. He was a Professor at the University of Illinois at Chicago from 1985–2018. He was also a visiting Professor at the University of Wisconsin, Madison; IMA, Minneapolis; IHES, Bures-sur-Yvette; IIT, Haifa; and Berlin Mathematical School. He has contributed to the fields of one complex variable, matrix and operator theory, numerical linear algebra, combinatorics, ergodic theory and dynamical systems, mathematical physics, mathematical biology, algebraic geometry, functional analysis, group theory, quantum theory, topological groups, and Lie groups. In his more than 200 publications he has solved many famous open problems, such as the notoriously "wild" problem of classifying tuples of matrices up to simultaneous similarity, Lax's eigenvalue crossing problem, and many others. Friedland has also contributed significantly to the theory of inverse eigenvalue problems and nonnegative matrices.

In addition, in 1978 he proved a conjecture of Erdos and Renyi on the permanent of doubly stochastic matrices. In 1993 he received the first Hans Schneider prize in Linear Algebra jointly with M. Fiedler and I. Gohberg, and in 2010 he was awarded a smoked salmon for solving the set-theoretic version of the salmon problem.

He has received many other numerous honors and awards for his contributions to mathematics, including a 2021 SIAM Fellow, "for deep and varied contributions to mathematics, especially linear algebra, matrix theory, and matrix computations." He also was elected a Fellow of the American Mathematical Society (Class of 2019).

Professor Friedland serves on the editorial boards of Electronic Journal of Linear Algebra and Linear Algebra and Its Applications, and from 1992 to 1997 he served on the editorial board of Random and Computational Dynamics. He is the author of the following books:

1. *Matrices: Algebra, Analysis and Applications* (World Scientific, 2015)
2. *Linear Algebra and Matrices* (joint with Mohsen Aliabadi, SIAM, 2018)

Mohsen Aliabadi received his Ph. D. in Pure Mathematics from the Department of Mathematics, Statistics, and Computer Science at the University of Illinois at Chicago under the supervision of Shmuel Friedland in 2020. He was a postdoctoral fellow at Iowa State University from 2020–2022 where he received Postdoctoral Scholar Excellence Award for Teaching and Mentoring Students, and Postdoctoral Scholar Research Excellence Award, in 2021 and 2022, respectively. He is currently a Stephen E. Warschawski Assistant Professor of Mathematics at University of California, San Diego.

His interests include combinatorial number theory, linear and multilinear algebra, graph theory, and field theory. He has authored fifteen papers and one book.

https://doi.org/10.1515/9783111337388-011

Index

https://doi.org/10.1515/9783111337388-012

www.ingramcontent.com/pod-product-compliance
Lightning Source LLC
Chambersburg PA
CBHW061415210326
41598CB00035B/6223